Predator Empire

Predator Empire

Drone Warfare and Full Spectrum Dominance

Ian G. R. Shaw

University of Minnesota Press
Minneapolis • London

Published by the University of Minnesota Press
111 Third Avenue South, Suite 290
Minneapolis, MN 55401-2520
http://www.upress.umn.edu

Printed in the United States of America on acid-free paper

21 20 19 18 17 16 10 9 8 7 6 5 4 3 2 1

Library of Congress Cataloging-in-Publication Data
Shaw, Ian G. R.
Predator empire : drone warfare and full spectrum dominance / Ian G. R. Shaw.
Minneapolis : University of Minnesota Press, [2016] | Includes bibliographical references and index.
Identifiers: LCCN 2015036890 | ISBN 978-0-8166-9473-0 (hc) | ISBN 978-0-8166-9474-7 (pb)
Subjects: LCSH: United States—Military policy. | Military art and science—Technological innovations—Social aspects. | Drone aircraft—United States—Social aspects. | Targeted killing—United States—Social aspects. | National security—United States. | Terrorism—Government policy—United States. | Hegemony—United States. | Social control. | Police—United States. | Technology and civilization.
Classification: LCC UA23 .S464 2016 | DDC 355/.033573—dc23
LC record available at http://lccn.loc.gov/2015036890

Contents

Acknowledgments

Many people have both directly and indirectly helped with this book. First, I am indebted to the University of Arizona, where I received my master's and PhD between 2006 and 2011. Under the beating Tucson sun and in the geography department of a strange building called Harvill, I was lucky to learn from John Paul Jones III (my adviser), Sallie Marston, Paul Robbins, and Marv Waterstone. They opened my eyes and provided me with the kind of education and friendship that is rare. To friends and fellow graduate students there—Jessie Clark, Jeff Garmany, Lawrence Hoffman, Jason Jurjevich, Tom Keasling, Vanessa Massaro, Jen McCormack, Jamie McEvoy, Jacob Miller, Jared Powell, Anne Ranek, Audra White, and Scott Whitlock—I give a hearty thanks for making my journey so enjoyable, memorable, and all too fleeting. Special thanks go to Majed Akhter and Katie Meehan for inspiring me to write this book as well as being amazing collaborators.

Thanks go to the Economic and Social Research Council in the United Kingdom, the University of Glasgow, and the Urban Studies Foundation for partially funding the research for this book. I am fortunate to have made wonderful friends at Glasgow's School of Geographical and Earth Sciences where I presently work, especially, Cheryl McGeachan, Caleb Johnston, Hester Parr, Jo Sharp, and Olivia Stevenson. They made writing in an office for fifteen months during 2014 and 2015 somewhat bearable. Thanks go to Jason Weidemann at the University of Minnesota Press for guiding me through the publishing process and to Mike Hanson, the copy editor of this book. Paul Dickson kindly donated his archive on the Vietnam War from the late 1960s and early 1970s, which proved invaluable for writing chapter 2.

I am grateful for a loving family, especially my mum and dad and my sister Zara, who sparked my interest in philosophy at a crucial moment. My biggest thanks go to my best friend, Lauren Fulton, whose sacrifices and support made this book possible in so many ways.

Introduction

Understanding Empire

The Dome

In 1979 the United States was rocked by soaring oil prices. The country faced one of the worst energy crises in memory. Nowhere was this felt more than the small Vermont town of Winooski, where another ice-cold winter threatened to chill its 7,500 residents to the bone. With temperatures under twenty below zero and snowfall over seventy-five inches, the cost of heating homes was proving worrisome. But a cunning plan was hatched. A group of city planners approached Mark Tigan, the city's director of community development. These entrepreneurs had an idea that could shelter townspeople from the blizzards and slash heating bills. "I didn't hear one organized voice against it," said Tigan, "since it meant that they'd never have to shovel snow again. They thought of it as their little piece of Tampa Bay."[1] The idea would be lauded and mocked in equal measure. Why not build a gigantic dome over the town? A bubbled utopia sealed from the frosty outside.

The Winooski dome would measure 1.3 square miles, stand at 250 meters high, and be constructed from crystal-clear plastic. Fresh air could be circulated by large intake fans, and the dome would be held aloft by air pressure slightly higher than outdoors. "I like to think of Winooski as a place where new ideas are thought up all the time," said Ken Meyers, president of the town's "Dome Club."[2] The town applied for $55,000 of federal money from the Department of Housing and Urban Development (HUD). The sphere attracted considerable national media attention. One Kentucky paper wrote, "Science fiction writers have predicted a future where people are forced to live underground like moles. Most people think that is pretty depressing. But living under a plastic parasol that can shut out bad weather, fallout and other unpleasantness doesn't sound all that bad, especially in New England."[3] A local teacher even penned an ode to the artificial bubble: "Dome over Winooski, / Not far from the lake; / Transparent and plastic, / Still real and not fake."[4]

In May 1980, after considerable excitement, HUD rejected the request. Despite widespread curiosity, many residents were fearful of the project. Who would clean it? Would life inside feel claustrophobic? Enthusiasm for the bubble didn't die immediately, however. The town hosted a dome symposium that attracted one thousand participants, including renowned inventor R. Buckminster Fuller, who had designed geodesic domes around the world. The next decade, another type of dome stirred the human imagination, only this time it was built for science. In September 1991 a crew of eight men and women stepped into Biosphere 2, a 3.15-acre research facility in Arizona. This enormous greenhouse, resembling a prototype for a space station on another planet, was engineered to create a series of self-sustaining ecosystems. These included rainforest, savannah, and even ocean biomes. The team of biospherians managed to survive in the dome for two years, despite fluctuations in oxygen levels, endemic hunger, social conflict (the group split into two factions), and an explosion in the ant and cockroach population.

The enthusiasm, fear, and curiosity for life inside these giant terrariums exemplifies a more basic architectural truth about humanity: we are builders of worlds. Our anthropology, our very human existence, is shaped by the artificial environments, big and small, we carve out from the planet. Hannah Arendt wrote, "Whatever touches or enters into a sustained relationship with human life immediately assumes the character of a condition of human existence."[5] Her point is as simple as it is radical: the human is constituted by the nonhuman.

Whether we build gigantic domes or robotic drones, artificial fabrications condition the spaces of human coexistence. Peter Sloterdijk defines these spaces of shared dwelling as "spheres." These spheres can be thought of as the biological, cultural, and technological enclosures that surround human beings. As he defines them, "The sphere is the interior, disclosed, shared realm inhabited by humans."[6] From the very first biological sphere, the mother's womb, to the artificial spheres of a mechanized civilization, we never stop passing through spaces that contain us, shape us, and transform who we are.

Spheres, in short, enclose human beings inside unique existential shells, constituting the rich kaleidoscope of our being. We are always "with" someone or something, and in this sense we always exist "outside" ourselves. As Judith Butler writes, "We are, as it were, social beings from the start, dependent on what is outside ourselves, on others, on institutions, and on sustained and sustainable environments, and so are, in this sense, precarious."[7]

In the modern age human coexistence has been subjected to increasingly technical forms of control. Modern living has provided both the conditions and the pathologies for the mass enclosure of life. From maximum-security prisons to CCTV on the streets to drones in the sky, we are constantly watched and surveilled, regardless of guilt or innocence. Millions of us have been herded into the *great civilizatory inside*. Life on planet earth is now time spent passing through enclosures of different sophistication and density. This is our *dome-estic life*. In 1981 the mayor of Winooski remarked, "I'm not a sociologist. But the idea of people living together in a controlled environment is a much more complex question than any of the technical concerns."[8] This complex question remains key to understanding this book: what does it mean to live on a planet that is enclosing its populations inside controlled, artificial, and dronified environments?

Beyond the Winooski dome, popular culture is full of domes: artificial skins grafted over human populations to form carceral shells. In *The Truman Show*, for example, the protagonist plays out his entire life inside a vast reality television show, with cameras hidden inside the carefully constructed set. Truman is locked in a daily routine he doesn't question until the artificial sphere that surrounds him begins to rupture. The film ends as Truman escapes his perfect home on a boat. Setting sail on what appears as a vast ocean, his vessel soon bumps into the edge of the dome, piercing its painted blue sky. Stephen King's *Under the Dome* depicts the slow descent into madness and civil war for one fictional Maine town after a dome seals the hapless residents inside. These fictional domes crystallize our anxiety, even curiosity, about life inside what are essentially prisons. The dome is, then, an architectural paradigm for the more general atmospheric enclosure of humanity.

The dome is also important for conceptualizing contemporary security. Not only does it embody a totalizing form of surveillance, but it also symbolizes the aerial dimension of state power. Today, the lower and upper atmospheres of the earth are swimming with satellites, airplanes, and drones. These machines transmit—across their antennas—telephone conversations between friends, soccer games between nations, directions to lost taxi drivers, and instructions for military assassinations in Yemen. State power in the twenty-first century is incredibly atmospheric.

The U.S. military defines *full spectrum dominance* as the control of all the physical domains of the earth—from the seas to the skies. Although the term is a well-known piece of jargon, it expresses the spheric dimension of human security more generally. To be safe and secure is to be housed within an enclosure, some kind of dome that protects its inhabitants from the outside. On the smallest of scales, this manifests itself in the alarmed suburban house or the gated community that keeps inhabitants protected from the "outside." On a bigger scale, the Reagan-era global missile shield, the so-called Star Wars system, was meant to protect the U.S. continent from Soviet missiles during the Cold War. Relatedly, consider Israel's more recent Iron Dome antirocket system. The dome is thus a cartographic design for an atmospheric form of security, an enclosure that protects against horizontal and vertical intrusions, putting a roof—electromagnetic or otherwise—above our heads.

And here lies the essential, provocative image: in the gap between humanity and the cosmos, a synthetic membrane is stretched across the planet—an artificial civilizatory world mediates and contains human existence. As will become clearer in this introduction, artificial or nonhuman sources of power are extremely important in the contemporary landscape of international relations. To an ever-greater extent, machines perform the atmospheric enclosure of the earth, manufacturing and policing spheres for us to live and die within, regulating the geopolitical climate of our everyday existence. To be human now means to-be-with-machines. As Arendt writes, "If the human condition consists in man's being a conditioned being for whom everything, given or man-made, immediately becomes a condition of his further existence, then man 'adjusted' himself to an environment of machines the moment he designed them. They certainly have become as inalienable a condition of our existence as tools and implements were in all previous ages."[9]

The machine of interest in this book is the drone. Drones, or unmanned aerial vehicles, are remotely piloted aircraft of various sizes and levels of sophistication, and they are transforming the geographies and infrastructures of state violence. If all objects mediate the human condition in some small way, then this raises bigger questions about how drones are changing the future of war, security, and freedom. These eyes in the sky are rewiring the international system, challenging the meaning of sovereignty, territory, and even domestic law enforcement. Accordingly, this book asks, what does it mean for humans to exist in an era of dronified state violence? To answer such a question, the book focuses on the case of U.S. drone warfare and argues we are witnessing a transition from a labor-intensive American empire to what this book calls a machine- or capital-intensive Predator Empire.[10]

At its core, then, this book is a provocative investigation into the geographies of U.S. drone warfare—one that is underwritten by a wider existential consideration of human being. Guiding these questions is an analytic based on a "more-than-human geopolitics," one that foregrounds the materiality of state power. The theoretical framework engineered throughout this book complicates the apparent division between the technical, the political, and the existential. Following philosophers such as Arendt, Sloterdijk, and Bernard Stiegler, I argue the nonhuman landscape—populated by objects, things, tools, and technology—directly influences the human condition. Or put another way, the infrastructures we build here on earth directly condition the spaces of everyday life, from the conduct of state violence down to our psychological dispositions. In this sense, drones are not simply bound to international relations or geopolitics but part of something much bigger.

Drones emerge not from a vacuum but from a history of human surveillance and warfare. So while drones are creating unprecedented forms of state violence and producing new geopolitical spaces, they nonetheless arise from preexisting conditions. It is therefore important to understand the drone as both a cause of geopolitical change and an effect of wider circuits of power and violence. For this reason U.S. drone warfare needs to be understood in terms of the growth of the U.S. national security state: the conglomeration of military institutions, intelligence agencies, and police organizations designed to protect the U.S. homeland.

The national security state can be traced back to the signing of the 1947 National Security Act by President Truman. This act created the CIA, the U.S. Air Force, and the National Military Establishment (renamed the Department of Defense in 1949) and formalized Cold War strategy. Moreover, the act began the long process of converting the social welfare state, nurtured under the New Deal, into a national security state, or what Tom Engelhardt calls the Fourth Branch of government.[11] Furthermore, the national security state is inseparable from a gigantic military–industrial complex and a national security economy. This has driven a militaristic foreign policy and a pervasive militaristic culture.

This book labels the present and future U.S. national security state as the Predator Empire. The Predator Empire is a concept used to gather together and theorize the multiple military, policing, and surveillance apparatuses that coordinate an increasingly dronified war on terror. More specific, the Predator Empire materializes a mode of state power (policing), a military strategy (predation), an archetypal technology of remote surveillance (the Predator and Predator B drone), and a geographical scale (the planetary). All of these—policing, predation, the Predator, and the planet—converge around the belief that the U.S. military is the premier guardian of civilization, a theme that has been a mainstay of the war on terror and, before that, the Cold War.

The approach this book takes is, therefore, to see drone warfare as a part of a wider project to surveil and enclose the human species. As such, it traces major spaces of human enclosure and argues that for decades—centuries even—human existence has slowly but surely been brought inside technological civilization (an idea explored in chapter 1). As will become clear, living inside technological civilization produces numerous contradictions that must be violently policed. So while the history of the drone is important (and covered in chapter 2), so too is the social war that rumbles across technological civilization, creating endemic forms of insecurity. If, as Sloterdijk argues, "there is no traditional empire that failed to secure its borders by cosmological means," then we can see in the Predator Empire a new state cosmology based on the enclosure of the sky with drones.[12] The drone must be understood as a mediator of state power, one that works to change the very terrain and logics upon which that power is subsequently exercised.

While *enclosure* has a very precise dictionary meaning—as a space that is bounded or fenced—it also expresses a much broader set of themes about historical acts of appropriation, confinement, and segregation. The enclosure of the commons, for example, was a period in English history, roughly between the sixteenth and nineteenth centuries, when common lands were privatized—or simply stolen—by a landowning class. In turn, agricultural communities were remade as strangers in an altogether strange environment. Enclosure was a kind of social war, and its victory depended on partitioning the once open field system with hedges and fences, creating segregated territories, or enclosures. In addition, the English government began creating disciplinary enclosures such as prisons, asylums, and poor houses to secure a growing mass of unemployed and alienated individuals. As Peter Linebaugh argues, "The incessant accumulation of 'industrial' subjects required their enclosure from the cradle to grave. To be ruled the population of civil society had to be confined and to be confined it had to be brought under complete surveillance."[13] Paradoxically, *state enclosures were built to mitigate the effect of state enclosure.*

This historical act of enclosure, a vital precursor to the Industrial Revolution, always had an important existential dimension. The privatization, division, and policing of the countryside wreaked havoc upon countless communities and birthed into the world an extremely atomized, securitized, and surveilled society. Since the dawn of the modern age, then, enclosure has been a project to bring the planet's inhabitants to the great *inside* of technological civilization: on the inside of its legal regime, on the inside of its economic system, on the inside of its architectural spheres, and on the inside of its surveillance apparatuses. With the passage of time, the apparatuses for enclosing the species have become only more atmospheric, more machinic, more militarized. Never before in human history has our globe been ensnared by so many surveillance apparatuses. Yet instead of simply enclosing people within physical architectures, the Predator Empire uses satellites, drones, and software algorithms to secure the spheres in which individuals are born, become, and die. We are not simply housed beneath the dome, then, but housed beneath the drone.

The Drone

Could robots one day erase humanity? This was the question University of Cambridge researchers began asking at the close of 2012 at the Centre for the Study of Existential Risk. One professor remarked that as technology escapes the constraints of biology humanity could find itself at the mercy of "machines that are not malicious, but machines whose interests don't include us."[14] That same month, a report by Human Rights Watch and the International Human Rights Clinic was released titled *Losing Humanity: The Case against Killer Robots*. It was the first publication of its kind and called for states to preemptively ban robots before they became fully autonomous and posed a threat to humanity.[15] And at the end of 2014, Professor Stephen Hawking warned, "The development of full artificial intelligence could spell the end of the human race."[16] Imagine, then, the danger posed by weaponized drones spilling into the skies with intelligence far beyond that of the human mind—a kind of *Terminator* planet.

These various reports and predictions are interesting. But must a drone, robot, or any other kind of machine possess advanced artificial intelligence to pose an existential risk? What if we imagine the existential risk more *broadly* as the capacity for objects and technologies to create and maintain unequal modes of power and domination? What if there is a risk that international relations are fast becoming the relations between machines as much as between humans? Like an electronic umbilical cord, U.S. foreign policy is bound to the infrastructures it has engineered in the pursuit of its war on terror. State power, in this sense, emerges from the complex exchanges between machines.

The U.S. military has been, and remains, a world leader in remote targeted killings. The drone has become central to U.S. national security strategy, which has switched from counterinsurgency in the city to counterterrorism from the skies. Whatever the size of the drone, they all essentially perform the same functions: intelligence, surveillance, and reconnaissance (ISR). From monitoring North Vietnamese troops in the 1960s to stalking al-Qaeda targets in Pakistan, the military has long desired eyes in the sky. Today, these eyes are becoming increasingly robotic and are part of a radical realignment of military power in the twenty-first century: the "dronification of state violence."[17]

The U.S. military's fleet of drones varies by size, shape, and sophistication, from the army's hand-thrown Ravens to the air force's Global Hawk, which can reach altitudes of sixty thousand feet. The year before the terrorist attacks of September 11, 2001, drone funding stood at around $284 million. By the fiscal year 2016, the Pentagon plans to spend close to $3 billion on drones. Indeed, between 2002 and 2010 the Pentagon's inventory of drones increased fortyfold, and it now owns a fleet of some eleven thousand drones, hundreds of which are weaponized.[18] The medium-sized Predator, for example, and its larger offspring, the Reaper (previously known as the Predator B), have both been workhorses of the U.S.-led war on terror. As of 2013 the U.S. Department of Defense had 237 Predators and 112 Reaper drones, both of which can be remotely piloted from across the planet. By the end of 2015, these hunter–killer drones had been used over five hundred times to kill an estimated 3,922 people outside traditional battlefields.[19]

With news that one hundred thousand troops were being cut from the U.S. military in 2012, it became clear the drone wars were here to stay.[20] "The world's concern," warns David Sanger, "is that the United States will use its technological advantage to create a new form of unilateralism."[21] And it is precisely this technological advantage that maintains the U.S. military's position as a global hegemon, or a Predator Empire. "With an agile force directed via a robotic information infrastructure," writes Alfred McCoy, "the United States could, in principle, parlay its military power into a second American century. If this interpretation is correct, then continuing technological advances could possibly exempt Washington from past patterns of imperial decline, creating something akin to an endless American empire."[22] As it turns out, the Obama doctrine looks an awful lot like the Rumsfeld doctrine that preceded it by less than a decade: a faith in technology, airpower, and networked communications.

Contemporary state and nonstate violence is difficult to contain. With recent U.S. air strikes across Iraq and Syria, together with a global ISR infrastructure, there is a sense in which *battlefield* is an archaic geographical term. The U.S. national security state is increasingly based on this borderless form of sovereignty, one that resonates with a range of extrajudicial spaces of control: from shadowy cyberwars against Iran to mass domestic surveillance by the National Security Agency (NSA) to expanding drone surveillance across Africa. Take Camp Lemonnier as an example. This is the U.S. military's only permanent base in Africa, located in Djibouti's main airport, and has served as a drone base for years (although nearby Chabelley Airfield now houses most drones). Thousands of U.S. special forces, civilians, and contractors have passed through this base as part of a strategy to target and eliminate Islamic militants in Yemen and Somalia and across the Sahel and Sahara. The twenty-first-century militarization of Africa is the latest phase in a globalizing U.S. security infrastructure that is converting the planet into a single battlespace and changing the geographies of state violence.

The Djibouti base is part of a growing network of military bases that house drones across the planet. "Even if the Pentagon budget were to shrink," writes Nick Turse, "the expansion of America's empire of drone bases is a sure thing in the years to come. Drones are the bedrock of Washington's future military planning and—with counterinsurgency out of favor—one of the preferred ways of carrying out attacks abroad."[23] While many of the gargantuan U.S. bases in Iraq and Afghanistan are now relics of the occupation, the U.S. military has been busy constructing smaller, more remote outposts across the world. These so-called lily pads aim to cover more of the planet with less of a footprint. The territorial extent of U.S. sovereignty has been significantly expanded by this strategy. The list of countries that have been surveilled by U.S. drones, or have housed U.S. drones, is long and includes Afghanistan, the Democratic Republic of Congo, Djibouti, Ethiopia, Iraq, Iran, Libya, Mali, Nigeria, Niger, Pakistan, Saudi Arabia, the Seychelles, Somalia, Syria, and Yemen. The drone atlas is constantly shifting. And back in the U.S homeland, the everyday spaces of human coexistence are being increasingly targeted by a dronified form of policing preoccupied with surveilling every facet of life.

In short, the new face of the U.S. military's empire has far fewer human faces. After the brutal counterinsurgencies of Iraq and Afghanistan, Washington's war managers scrambled to find a new strategy. Their response was to do more with less as robots, drones, and satellites began to redistribute and replace human bodies (and therefore vulnerabilities): shifting personnel from the frontline and putting them in service of their robotic proxies. In other words, American empire is transforming from a labor-intensive to a machine- or capital-intensive system: the Predator Empire.

A vast policing infrastructure of machines, special forces, drones, and software algorithms hunt for threats across a totalizing battlespace that connects the homeland with the battlefield. The geopolitics pursued by the Predator Empire is different, however, from those of either the Roman or the British empire. Rather than commanding and controlling the physical landscape, aerial technologies now hack the human lifeworld from the skies in order to capture, digitize, and police it. Here, the drone acts like a virus: boring into the existential shells of human beings, reprogramming the climate of their interiors. "All objects," argues Graham Harman, "constitute their surroundings retroactively—objects are *retroviruses,* injecting their own DNA back in the nucleus of everything they encounter. It is not just humans who do this."[24] Under a system of aerial surveillance and assassination, securing territory is secondary to capturing, coding, and reprogramming human life. This means that technology must be understood as an existential force, since it changes our relationship with the world, with the state, and with each other. Without this understanding, we will fail to grasp the real existential risk drone warfare heralds for the future of international relations, domestic law enforcement, and the everyday spaces of life, justice, and liberty.

Military and nonmilitary technologies have consistently rewired the conduct of state power in profound ways. Telegraphs, railways, tanks, aircraft, nuclear weapons, drones, and the Internet have all brought with them distinct geographies of violence and control. In this sense, *geopolitics is overrun by technological forces*: electronic infrastructures rewire international relations in profound ways. So while war may be influenced by all kinds of discourses, opinions, strategies, and tactics, for Jacques Ellul the situation is clear: "One factor always upsets everything: the machine."[25] Government policy is constantly playing catch-up with the ingenuity of the military–industrial complex and its machinic creations. Indeed, in less than a decade targeted killings delivered by Predator drones have become so normalized that the Obama administration has constructed a streamlined bureaucratic database for administering death. The disposition matrix contains an electronic list of suspects targeted for elimination across the planet.

Many scholars writing about drones mobilize their critiques around international law or ethics. Often, the concern is with how drones are producing a detached, video game style of war. The growing use of drones, warns Peter Singer, could transform the "public into the equivalent of sports fans watching war, rather than citizens sharing in its importance."[26] But such concerns about the aesthetics of killing leave untouched the dominion of humans over their technological creations. Drones are viewed as tools of state power, used by rational actors for rational ends. More generally, the problem with the traditional academic field of international relations (IR), according to Harman, is that "the only ordering agent on the scene is *people,* in the form of the sovereign human or humans. In other words, the problem with this model is that it focuses on humans generally, and the state specifically, as the privileged sites of political order."[27] But objects can soon come to master the masters. As Gabrielle Hecht writes, "Material things can be more flexible—and more unpredictable—than their builders realize."[28] Whoever sits in the White House in the future will inherit a technological apparatus beyond their direct control (and, indeed, beyond anybody's direct control).

This book therefore goes beyond contemporary analyses in IR and political geography to foreground the existential transformations created by living and dying in a booming Droneworld. Rather than being understood as instruments, drones are seen as geopolitical agents creating new modes of state power. The materiality matters. Of course, humans are important to the exercise of state power. But there are "unknown unknowns" set in motion with the birth of every machine. Technologies can rebel and transform their surrounding world, like a virus that reprograms the cells of its unwitting host. A sprawling set of surveillance apparatuses—roaming across land, sea, and outer space—now leave few places left on earth to hide. Drone warfare, in short, requires us to seek new ways to understand empire.

"Empires decline and disappear," argues McCoy. "But empire in some form has persisted over the millennia, and will likely continue into the foreseeable future."[29] Past empires, such as the Roman and British empires, mobilized technology to their advantage, whether road or railway. Indeed, empires would be impossible without the infrastructures that anchor their power relations to the landscape. The Predator Empire is no different. It exercises power through the hyperconnected infrastructures, machines, and bodies that bind us together on earth. In order to strike a member of al-Shabab in Somalia, for example, the U.S. military relies on a drone base in Djibouti, satellites in outer space, a command hub in Germany, and commercial fiber optic cables that snake across Europe and the Atlantic floor. This global coimbrication means, in turn, that any unforeseen consequences of the Predator Empire belong to all of us. "Because we live in an increasingly interconnected international system," argues Chalmers Johnson, "we are all, in a sense, living in a blowback world."[30] Blowback is bigger than any nation, and it is bigger than any military. American or not, we are all subjects of the titanic clash between life and those technologies that seek to enclose it.

Understanding Empire: The Leviathan

The Leviathan is a mythical beast depicted throughout ancient history. Variously represented as a whale, a dragon, and a sea serpent, the monster appears in the book of Job, which describes the Leviathan as a “creature without fear,” a terrifying “king over all that are proud.” This Old Testament passage portrays Job as a hapless figure trembling before the invulnerable Leviathan. “If you lay a hand on it, you will remember the struggle and never do it again! Any hope of subduing it is false; the mere sight of it is overpowering.” The symbolism is certainly provocative: humanity is plagued and overawed by forces far bigger than itself. *Leviathan* is also the title of Thomas Hobbes’s monumental work of political philosophy from the seventeenth century.[31]

Written in the midst of the English Civil War, Hobbes’s treatise on sovereignty assumes the worst traits of humanity. Hobbes believes that human nature is defined by a ruinous instinct of self-preservation. Individuals will do anything to ensure their own survival. There is no natural law other than this brute survivalism, and since each individual has an equal capacity to kill, everybody is at risk of death from his or her neighbor. Paradoxically, it is equality that leads to endemic danger. This so-called state of nature is what Hobbes defines as a “war of all against all” in which there is “continual fear, and danger of violent death, and the life of man, solitary, nasty, brutish, and short.”[32] To escape this misery, humanity must trade its freedom for sovereign protection. In this exchange of liberty for the rule of law—the foundational social contract—the Leviathan is born. This is a commonwealth that binds the bodies of every man, woman, and child together and is headed by the sovereign (either a monarch or a parliament).

Hobbes’s Leviathan is a uniquely human creation, embodying a conglomerate of people. This construction inaugurated a trajectory in political philosophy that assumes power and politics are purely human activities. Indeed, in international relations today the idea that it is people who are in control is widely accepted.[33] But this book thinks otherwise. Rather than injecting power solely in humans, it asks how power is distributed among the objects, technologies, and infrastructures that bind humanity together on earth. As Bruno Latour asks, “Where has political philosophy turned its distracted gaze while so many objects were drawn under its very nose?”[34] We must, therefore, employ a more-than-human geopolitics to fish for another type of monster in our exploration of U.S. drone warfare, one that is assembled by satellites and drones as much as by kings and queens. It is with this materialist outlook that this book explores the causes, conditions, and consequences of drone warfare, moving past mainstream accounts to hook a different kind of Leviathan: the Predator Empire.

State power must be built, constructed, and engineered: it is less a resource embodied in people than a type of scaffolding grafted across the planet. This is one of the core insights from actor-network theory (ANT), which aims to correct the modern assumption that objects and things are puppets of human masters. "It's not unfair to say that political philosophy has often been the victim of a strong object-avoidance tendency."[35] Likewise, the division between categories like *society* and *nature* is criticized by ANT approaches and is replaced with the idea of material *actors* (human and nonhuman) producing and moving through various networks. Typically, however, this is not a starting point in IR scholarship, which begins with an assumption similar to Hobbes's: both the state of nature and the Leviathan are made of people. IR is human, all too human. One immediate problem with this realist attitude is that any analysis of drone warfare—whether hawkish ("drones are vital to national security"), hopeful ("drones play a vital humanitarian role"), instrumental ("drones are just tools"), or legalistic ("drones break or enforce international law")—misses the existential dimension of how drones change the ways in which we relate to the state and to each other, not to mention their future uses and abuses. The drone must be seen, therefore, as a geopolitical actor.

Whenever IR does invite technology to the table, scholars typically understand it narrowly. For example, while technology can change the conduct of military violence, it remains *exogenous* to society.[36] The same critique can be made of geopolitics, even if it has historically included the materiality of the planet in its foreign policy calculations. The point, in either case, is that there is no "outside" of the social that, in turn, interrupts the social. It is this prima facie modernist division that leads to all kinds of confusion. As Nick Srnicek concludes, despite claims to the contrary, mainstream international relations "is imbibed with a thorough immaterialism. Disembodied actors, interests, intentionality, and instrumental rationality are the substance of much IR."[37] So long as geography, technology, and other material infrastructures are understood as passive backgrounds or mere conductors of power, the scope of drone warfare is narrowed to a form of humanism that underplays its potential impact. We must redistribute agency—and geopolitics—across a global network of things.

Another problem with mainstream international relations that must be tackled is its scalar assumptions. The globe is often understood as a giant container that, like a matryoshka doll, holds smaller spaces inside it: the international, the national, the regional, and the local. Power then becomes a question of hierarchical authority. In other words, IR usually assumes multiple layers of reality. And yet as the work of political geography has repeatedly shown, scale can be misleading, since it presupposes a bounded form of organization that does not always reflect the order of things.[38] Now more than ever, territory and sovereignty do not straightforwardly overlap. As the planet has become more globalized, its mosaic of distinct states has melted into a more distorted painting: a scattershot of statelets, militarized cities, and transnational flows. In other words, the international and the domestic are not straightforwardly recognizable.

In short, classical social contract theory assumes that both sovereignty and territory are human creations. How, then, do we extricate ourselves from this predicament? Do we abandon the model of the Leviathan? For Bruno Latour and Michael Callon, the problematic posed by Hobbes's Leviathan, despite its anthropocentric construction, remains central to thinking about state power. Rather than view the Leviathan as a mangle of human bodies, however, they see the Leviathan as a kind of macroactor constituted by a multitude of human and nonhuman actors. An actor is here defined as "any element which bends space around itself, makes other elements dependent upon itself and translates their will into a language of its own."[39] The social contract posed by Hobbes, rather than being a legalistic construct, is better understood as one instance of "translation," whereby humans are corralled into certain habits, dispositions, and actions by the anchoring power of nonhuman elements. "In order to build the Leviathan it is necessary to enroll a *little more* than relationships, alliances and friendships."[40] One must enroll a whole manner of objects to execute state power.

Crucial here is the insight that humans alone cannot stabilize the Leviathan: state power must be translated into objects and technologies more durable than our fleeting human lives. As Latour and Callon write, "But if you transform the state of nature, replacing unsettled alliances as much as you can with walls and written contracts, the ranks with uniforms and tattoos and reversible friendships with names and signs, then you will obtain the Leviathan."[41]

No matter how powerful kings and queens may have once appeared, no social contract is as awe inspiring as the materials, tools, and objects that enclose humanity in monstrous formations. State power accumulates through the aggregation—and stabilization—of multiple actors and networks. It is size rather than scale that is crucial to the asymmetrical geopolitical landscape. As Srnicek argues, "From this it can be concluded that a minimal condition for being global is the capacity to affect large numbers of actors that are widely dispersed throughout a series of assemblages."[42] Networks of force relations, in other words, do not simply rest upon social relations. And neither does the Predator Empire.

If Latour's state of nature is the primordial world of unmediated interactions—of naked humanity—then the Leviathan is the state of mediated interactions. And if the Leviathan is a monster, it's because of its hybrid, cyborgian fusion of body parts, tools, minds, and machines. "The Leviathan is monstrous too because Hobbes built it using only contracts and the bodies of ideal, supposedly naked, men. But since the actors triumph by associating with themselves other elements than the bodies of men, the result is terrifying."[43] The ANT Leviathan is an alliance of humans and nonhumans and remains a guiding model for the Predator Empire. As Peer Schouten writes, "Where critical approaches to IR theory consider the state of nature a *cultural* construct, ANT retorts that political society is a *socio-material* (or material-semiotic) construction."[44] To put it bluntly: power is mediation, and the modern social contract is really an unending negotiation with nonhuman apparatuses.

For Hobbes the birth of the Leviathan is meant to end the anarchic, individualized violence of primordial humanity. Yet, as Latour replies, "*we have never left the state of war,* the *state of nature* that Hobbes thought the Leviathan had gotten us out of."[45] This is because translation never stops running as the background condition of human association: a war of mediation continually translates how humans interact with each other and with the state. To be human is to be endlessly reconfigured with nonhuman elements. Given that political stabilization today rests upon the work of nonhuman actors even more than human ones, the problematic posed by the Predator Empire is doubly important. As this book will go on to argue, the Leviathan's enclosure of human life does not so much end violence as it institutionalizes and stabilizes a pervasive social war. If, as Andrew Barry warns, "arguably, actor-network theory had over-expanded the notion of politics, thereby forgetting any sense of the specificity of what is conventionally understood to be politics," then this book fuses a more-than-human geopolitics to the idea of a social war to better elaborate the ongoing discontents of sovereignty.[46]

The War of All against All?

There is another sense in which Hobbes's "war of all against all" is tied more obviously to this book's central thematic: the spread of U.S. drone warfare across the planet. Mainstream IR globalizes Hobbes's idea of a state of nature into an anarchical international system. As the assumption goes, the entire planet exists in a lawless war of all against all that must be sliced up and subdued by capable sovereigns, each defending their own patch of territory in the world system. These realist assumptions about human nature, international anarchy, and the (absolute) agency of the sovereign remain a mainstay of foreign policy. And yet, more than ever, in the fourteenth year of the U.S. war on terror, what Derek Gregory has called an "everywhere war"[47] continues to erupt across the globe: from the spread of the so-called Islamic State in Iraq and Syria to al-Qaeda affiliates in North Africa. As Michael Hardt and Antonio Negri write, "Today it is increasingly difficult for the ideologues of the United States to name a single, unified enemy; rather, there seem to be minor and elusive enemies everywhere. The end of the crisis of modernity has given rise to a proliferation of minor and indefinite crises . . . an omni-crisis."[48] The rise of powerful nonstate actors, fragmented forms of violence, and porous national boundaries continues to reconfigure the conduct and spatiality of global conflict.

Since at least the Bosnia-Herzegovina conflict between 1992 and 1995, warfare has shifted from primarily state-oriented violence, involving a mass of soldiers and vertical hierarchies, to a series of hybrid or "low-intensity" conflicts that involve private contractors, paramilitaries, and criminals. So argues Mary Kaldor, who describes the rise of "new wars," defined by a clash of religious and ethnic identities and underwritten by a global war economy.[49] What we imagine as an old war originated in Europe between the fifteenth and eighteenth centuries with the rise of the nation-state. War was understood as a discrete, large-scale event between sovereign nations, often fought over territory and waged by distinguishable combatants. The geopolitical situation today is more fragmented and complex. War, if that term has any meaning, has become "the pervasive *matrix* within which social life is constituted," and the battlefield has been replaced "by the multi-scalar, multidimensional '*battlespace*,'" in Gregory's words.[50]

The implication of both Kaldor's and Gregory's argument is that it is no longer possible to contain war to a single space: the battlefields of old have morphed into an emergent battlespace in which even the idea of a soldier with a national identity is changing. For example, in 2015 the United Nations reported that 25,000 individuals from over one hundred nations had joined al-Qaeda and Islamic State affiliates, the vast majority concentrated in Iraq and Syria. Zones of conflict and peace are intermingling in overlapping spaces, producing contested statelets and ad hoc violent urban insurgencies with multiple actors and factions. "Everywhere, boundaries are being drawn between protected and prosperous global enclaves and anarchic, chaotic, poverty-stricken areas beyond," insists Kaldor.[51] The world is becoming more fragmented even as global forces glue us all together. Globalization must be understood in this most paradoxical of lights: as a universalizing space of disconnection. In its most radical form, this electronic gathering of millions of souls on the planet enables the pinpoint, near instantaneous assassination of single individuals.

Crucial to this omnicrisis is the so-called individualization of warfare. It is individuals rather than nation-states and their armies that are now targeted by the U.S. military. This process has accelerated with the rise of unmanned aircraft such as Predator and Reaper drones. These drones target "high-value" terrorists across an expanding battlespace. A predator is typically defined as an animal that preys on or devours other animals. In the years that have followed the terrorist attacks of September 11, 2001, U.S. national security strategy has come to resemble a form of predation: a manhunt directed against individual targets dotted across the globe. Predator and Reaper drones, together with special forces and NSA software algorithms, have been deployed against a growing list of international suspects. Harold Koh, the formal legal advisor to the U.S. Department of State, confided to a friend about targeted killings by drone, "I kept slipping back and forth between the view of the predator and the view of the prey."[52]

The drone war in the shadows of the globe is the antipode of the visceral, street-level counterinsurgencies pursued by the Bush administration: precise, economical, and deniable. But Obama's embrace of the controversial Bush-era doctrine of preemption has remained. On September 14, 2001, Bush defined this vision as "identifying and destroying the threat before it reaches our borders."[53] Aerial assassination adheres to this same logic. And yet, as David Sanger explains, the strategy during the Obama era was "not preemption against a state, which was the logic of the Iraq invasion."[54] Instead, the president directed "preemptive strikes against terrorists who had struck before or who, intelligence showed, were suspected of planning attacks."[55] The spatial scale of the terrorist hunt has switched from the nation-state to the individual and is delivered by machine rather than marine. Driving this manhunt for the past decade has been the drone itself: a technology capable of producing overlapping spheres of real-time surveillance imagery to politicians sitting thousands of miles away.

The drone both crystallizes and galvanizes the trend toward what Pentagon officials call "manhunting." The manhunt can take place over continents, and the idea of a national border is anathema to the Predator Empire. What this book makes clear is that the traditional idea of war no longer captures the profound social unrest that bubbles across the planet. In this sense no-knock SWAT raids in Los Angeles connect with drone strikes in Pakistan's tribal areas. Both are part of a continuum of violence this book conceptualizes as the Predator Empire. The use of enclosure remains a core concept throughout the book, since it bridges the idea of military conflict and surveillance to more systematic forms of social violence caused by living inside an increasingly enclosed technological civilization. What follows is an outline of the major themes in the remaining chapters of this book.

The Long March of Human Enclosure

An important relationship explored throughout this book is that between humans and machines. How do nonhumans interact with, as well as transform, human existence and international relations? The idea of technological civilization outlined in chapter 1 expands and consolidates many of the theoretical issues raised in this introduction. The modern age is, above all, defined by the embedding of technology in the most intimate and diffuse spaces of the planet. Technology has enabled societies to become ever more complex through nonhuman forms of mediation (which have simultaneously rendered more and more people economically superfluous). As such, chapter 1 argues that the concept of technological civilization best encapsulates our dense technical living. Technological civilization is constituted by the artificial, increasingly capital-intensive infrastructures that enclose humanity.

Of course, humans have always employed objects, tools, and machines in their social lives. As Latour argues, human society is nothing without nonhuman mediators. So why use such a grand label as *technological civilization* in this book? Because we are reaching a point in which mediation is now *predominantly* nonhuman. This is why to speak of technological civilization is to outline a form of social organization and state power that is historically far more artificially mediated than previous human constellations. This idea therefore brings to the surface a much deeper meditation on the specific role of technology in mediating human subjectivity in a globalized world. These theoretical issues and debates will remain present throughout the book as they wind their way through the Vietnam War, the war on terror, NSA surveillance, and militarized policing.

The traditional idea of American empire thus needs to be rethought in light of the proliferation of intelligent and increasingly autonomous nonhuman actors. The theoretical framework provided by a more-than-human geopolitics goes some way to addressing this need for an understanding of the role of materiality in the execution of state power and violence. As such, a conceptual exploration of objects, technology, and international relations is put forward. In turn, U.S. drone warfare must be situated within a longer historicoexistential trajectory of enclosure. This begins with an examination of the historical act of enclosure in England, which was a long period—from around the sixteenth to the nineteenth century—of intense agricultural transformation that cleared people from the land and produced vast numbers of surplus populations.

Enclosure as a philosophy, practice, and logic of domination was exported across the world system during colonial times. Indeed, it goes hand in hand with what this book calls *imperial space,* which is the homogenizing, calculative, and ultimately destructive spatial order that European empires installed across the planet. The penultimate section of chapter 1 narrows its focus to the machine–human relationship and puts the work of Karl Marx, Gilbert Simondon, and Bernard Stiegler into dialogue. Machines are framed as disciplinary technologies that regulate the spheres of human coexistence. In this sense, "to be" is not simply *to be with* machines but *to be dominated by* machines. Technological civilization thus increasingly resembles an open prison.

The Vietnam War and Electronic Enclosure

Drone warfare, manhunting, and atmospheric forms of state power cannot be understood without investigating the legacies of the Vietnam War, which is the topic of chapter 2. In the decade of technologically intensive violence pursued by the U.S. military, which James Gibson labeled a "technowar,"[56] a key concept was advanced: the electronic battlefield. It was created in the 1960s by the Department of Defense to automate the battlefield with sensors, computers, and bombers. As individuals moved through the jungle, they triggered acoustic and seismic sensors that relayed radio signals to a U.S. military base in Thailand. Here, IBM supercomputers converted the signal into moving targets on-screen. The spatial coordinates were then sent to air force planes, which automatically bombed the location. This was "a new philosophy of war," explains Paul Dickson, "a manless, foolproof, giant, lethal pinball machine out of which no living thing could presumably escape."[57] This advanced technogeography, which came to include an unprecedented deployment of drones, spiraled beyond Southeast Asia and became a defining moment in the so-called revolution in military affairs.

In addition to creating the electronic battlefield, the U.S. military targeted the landscape and atmosphere more directly. An armada of machines and weapons modified the fields and jungles of South Vietnam, Laos, and Cambodia. A mixture of Agent Orange, napalm, bombs, and bulldozers was used to modify the Vietnamese landscape to fit the abstract blueprints of the war managers. The Vietnam War was a war against life that did not always differentiate between the human and the nonhuman—a war that used fire, tractors, and defoliants to punish, pacify, and eradicate entire ecosystems. This logic of atmospheric power extended to the smaller-scale use of tear gas, which soon boomeranged home to the protestors at Berkeley in 1969. Indeed, the Vietnam War was a major period of war coming home. For example, technologies from the electronic battlefield were deployed along the U.S.–Mexico border.

The Phoenix Program is the third legacy of the Vietnam War

crucial to the development of the Predator Empire. This was a type of manhunt pursued by U.S. and South Vietnamese special forces, under the supervision of the CIA, in which hunter–killer teams captured, interrogated, and at times killed what was known at the time as the Viet Cong Infrastructure. Horrendous acts were sometimes committed in a largely underreported, but highly controversial, U.S.-led operation beginning in the early 1960s. "Indeed, the Phoenix Program was set up by Americans on American assumptions, in support of American policies," according to Douglas Valentine.[58] Forty-six thousand were killed with little actionable intelligence.[59] Its grim legacy would come full circle in the manhunt pursued in Afghanistan and Iraq decades later.

Finally, the Vietnam War was a pivotal moment in the history of unmanned warfare, which saw the jet-powered Firebee drone become an aerial surveillance technology used across the denied airspaces of North Vietnam. There are five overlapping phases in this wider history: (1) the drone as a target, (2) the drone as a flying bomb, (3) the drone as a surveillance platform, (4) the drone as a hunter–killer, and most recently (5) the drone as a policing technology. In short, chapter 2 charts how the electronic battlefield and early experiments in drone surveillance, together with atmospheric warfare and manhunting, became central pillars of the modern Predator Empire.

Full Spectrum Dominance

Full spectrum dominance is a military concept—or ambition—for the total occupation and control of land, sea, outer space, cyberspace, and, even, psychological space. Chapter 3 explores how the world is becoming a seamless electronic battlespace and asks, How did the Predator Empire become a global condition? The Cold War was a crucial period in this expansionism, with the manhunt doctrine exported from Vietnam to South America by the CIA in the 1970s and 1980s. Focus then shifts toward the CIA's manhunt for Osama bin Laden in Afghanistan in the 1990s to the war on terror. During the Bush administration, special forces slowly eclipsed the CIA and became the U.S. military's principal counterterrorist force across the planet. In Iraq and Afghanistan, special forces would become an industrial killing and capturing machine.

Beyond declared theaters of war, the tribal areas of Pakistan, much more so than any other region, have been targeted by the CIA's "covert" Predator warfare. This form of targeted killing is rooted in the British colonial practice of air policing from the 1920s. Decades later, as the CIA's Predators struck al-Qaeda militants in the same region, the U.S. military's focus subsequently shifted to al-Qaeda offshoots, including the Islamic State in Iraq and Syria, al-Shabab in Somalia, al-Qaeda in the Arabian Peninsula, and more dispersed threats across the Maghreb and Sahara regions. Indeed, an increased—although uneven—militarization of Africa is apparent, with a variety of U.S. surveillance operations expanding across the continent.

This militarization of the planet would not be possible without an extensive network of bases built during the Cold War, which network together to produce the infrastructural skeleton of the Predator Empire. But as what Johnson labeled Baseworld has become a fixture of the planet, its geographical logics have shifted in recent years with the rise of remote warfare.[60] Drones are reengineering the spatiality of U.S. surveillance, converting Baseworld into the leaner but no less expansive Droneworld.

This expanding Droneworld is transforming the U.S. Navy. Although the military's manhunt will continue to be underwritten by a land-based infrastructure, the ocean will serve increasingly as a platform for unmanned wars. Ships in international waters radically extend U.S. sovereignty without the impediment of violating the territory of other states. Aircraft carriers, in particular, enable U.S. airpower to move in denied areas of the world. Moreover, the role of aircraft carriers is being reimagined. Smaller drones enable smaller carriers, which in turn can produce wider spheres of surveillance. The end goal is a distributed maritime system of drones that can be launched anywhere, anytime.

Although the land and sea are important domains for military domination, the final frontier of the Predator Empire is outer space. The occupation of outer space is vital to the Predator Empire's ongoing existence, since it enables communication, surveillance, geospatial intelligence, and of course, the remote-split operations of drone warfare. The enclosure of the earth's upper atmosphere began at the start of the Cold War with research into intercontinental ballistic missiles (ICBMs). President Ronald Reagan crystallized this arc of development with his so-called Star Wars speech in 1983, which called for a global "space shield." By the late 1990s the idea of space war had gained traction in military minds. No other country has antisatellite or orbital weapons in space, yet for U.S. galactic warriors this most daring form of militarization is viewed as an inevitable step in extraterrestrial national security.

The Rule by Nobody

Cyberspace is a crucial domain for full spectrum dominance and enables the kind of bulk surveillance at the heart of the Predator Empire. Guided by the philosophy of Hannah Arendt, chapter 4 investigates the role of machines in producing totalitarian systems of control. It begins by theorizing the Predator Empire as a type of biopolitical immune system. *Biopolitics* is a term used by Michel Foucault to describe a new regime of government power that targets the life of the species, or what he terms "state control of the biological."[61] Our species has never been so alienated and so connected, so insecure and so secured. If technological civilization fails to control its own irrational tendencies, its very existence is threatened. It must therefore develop *immune* capacities to quell systematic unrest.

Totalitarianism is centered, as Roberto Esposito argues, on this practice of protecting life against danger and is therefore a form of biopolitics that he calls "immunization."[62] Understood in this way, totalitarianism is the mass production of spaces of immunity. Individuals and communities are brought, willingly or unwillingly, inside secured state enclosures. The U.S. war on terror has sought to globalize such spaces of immunity, exporting the logic of a pacified homeland everywhere. The Predator Empire hacks into the global lifeworld in the name of immunity, restricting and modulating the patterns of human activity. The preservation of technological civilization, in short, reveals its hostile, even totalitarian obverse.

Today, a hyperrational form of bureaucratic authority governs technological civilization. Arendt called this a "rule by Nobody," an abstract system of control, a "tyranny without a tyrant," in which Nobody is held accountable for their actions.[63] With the sophisticated machines that enclose the planet, there is a sense that the rule by Nobody is fundamentally a rule by technics. Rationalized, mechanized, and automated killing lies at the heart of the U.S. planetary manhunt.

Indeed, computers now govern and discipline much of our speaking, acting, and thinking, in what Gilles Deleuze argues are contemporary "control societies."[64] In addition to enclosing land, sea, and space, then, the Predator Empire tracks our digital lives. A universal modulation resonates across the control societies of technological civilization, generating a widespread conformity. Human subjectivity is formed at the intersection of vast media infrastructures that work to appropriate the psychic interior of the individual. In this sense, the formation of the human psyche is a crucial site of biopolitical control.

The importance of computers in modulating state power reveals the centrality of the Internet in mediating technological civilization. At one end of the scale lies cyberwar and the direct militarization of the Internet. But more generally, digital enclosure can be thought of as the control of information that passes across the Internet. Software algorithms are, in particular, a vital part of U.S. national security and are used to identify and track suspicious patterns of life across enormous databases. Combined with the ability to digitize the planet through geospatial intelligence, which is handled by the National Geospatial Agency, the Predator Empire seeks to dominate the electromagnetic spectrum. Whereas empires have conquered and colonized territory for millennia, the Predator Empire seeks to surveil the global lifeworld. As McCoy writes, this is a "nonterritorial American imperium."[65] Distant spaces collapse together inside a computational topology.

The NSA is the agency most directly responsible for surveilling the planet's communications. The U.S. military leaned heavily on the NSA for its counterterrorist operations in Iraq and Afghanistan, and the agency remains a vital source of signals intelligence for drone operations and targeted killings across the globe. Privacy in such times is rapidly evaporating. The possibility that we can be watched by unseen authorities creates an unavoidable and creeping paranoia, which risks further corroding the spaces of liberty and free thinking. In sum, the Predator Empire imposes a rule by Nobody: a rule generated by thousands of apparatuses that hack, watch, and modulate the global lifeworld.

Policing Everything

What happens when the Predator Empire returns to the homeland? Indeed, what happens if it never left? Chapter 5 traces a wider, more foundational narrative that argues policing spirals beyond any particular institution and is instead a process of social discipline necessary to maintain the inequality at the heart of technological civilization. The functioning of civilization relies upon an endless internal pacification. In the age of electronic globalization, this social war is becoming a planetary condition. Everyone and everything must be kept in their right place.

The main empirical investigation pursued is the past, present, and future of U.S. policing. The war on crime, the war on drugs, and, later, the war on terror all provided the rhetorical gloss, resources, and legislation for the widespread militarization of the police. In such times, the everyday streets of America are viewed as battlefields, and civilians are treated like enemies of the state. In addition to evidence of widespread militarization, U.S. law enforcement has become increasingly preemptive, which is to say centered on predicting criminal activity in the future. This combines CCTV, geospatial intelligence, population databases, and algorithmic software to secure individuals and territories that may present a danger in the future. This reflects broader trends in the U.S. military, which is fixated on preemptive manhunting. The rush to secure time collapses the homeland and the battlefield within a single electronic battlespace.

The modern age's foundational social war was of course the enclosure of the commons. In the name of civilization, landowners threw commoners off the soil to "improve" England. Families were forced into the depraved conditions of industrializing towns. They lost their world but found civilization. In other words, enclosure is the originary act of worldly alienation: the accumulation of capital relied on the accumulation of worldless people. As E. P. Thompson wrote, "Enclosure (when all the sophistications are allowed for) was a plain enough case of class robbery."[66] Sovereignty, as such, needs to be analyzed, as Foucault argues, "in terms of the unending movement—which has no historical end—of the shifting relations that make some dominant over others."[67]

As alienation and unrest became systematic products of enclosure, the ruling classes scrambled to put a lid on the contradictions of their own creation—namely, the creation of a vast surplus population. An archipelago of what Foucault calls "disciplinary institutions" began to spring up, including prisons, factories, hospitals, schools, workhouses, and asylums. Hundreds and thousands of state enclosures began caging masses of homeless, surplus, and maladjusted people together. And while discipline was first contained in these state enclosures, it soon escaped from these institutions to enclose society under a generalizable mechanism of surveillance.

After the commons were enclosed, the archetypal space of human coexistence became the city, which has since become a battleground in the war on terror. "If the point of the war against terrorism is to pursue the enemy into his sociological and cultural labyrinth," argues Mike Davis, "then the poor peripheries of developing cities will be the permanent battlefields of the twenty-first century."[68] Indeed, the lives of the poor and the rich are being reorganized such that the two masses of humanity are no longer required to interact with each other. Elites plug into artificial, securitized cocoons, transferring bodies, minds, and capital into a network of exclusive insides fortified against the surplus masses on the outside. A new kind of ultrasecure existential anthropology has been forged in the contemporary metropolis.

Additionally, the atmosphere that surrounds urban environments is being secured. State power has ascended into the atmospheres that envelop us. The drone is a crucial machine for realizing this atmospheric policing. The final step to enclose technological civilization is, therefore, to saturate the air around us with swarms of microdrones: a militaristic dream of automated and roboticized policing in the city. The U.S. military envisions tiny drones—Nanos—interacting with each other in marauding clouds. The danger is, as Johnson warns, that "even an empire cannot control the long-term effects of its policies."[69] The robots in distant battlefields always return home. The endgame of the Predator Empire is a drone society that will mirror the imperial battlefields abroad, with police drones endlessly monitoring the public. The idea of a dronepolis, or a city of the drone, provides a blueprint for thinking through the dronified urban environments of the future.

Recapping the major themes of this book, the conclusion explores the continuing passage from the U.S. social security state to the national security state and how this new Leviathan—a mechanical monster stripped of its human flesh—is creating an unaccountable and alienated form of government violence. The conclusion is therefore the most important part of the book, since it extends its most significant arguments and reflects on the kind of future—a war of all against all—into which we are sleepwalking. If in the modern age humanity increasingly dwells in artificial shells of mediated existence, then consider the Predator Empire a universalizing shell shock: a rapidly expanding form of security—and trauma—tearing through the spheres of technological civilization. There is a profound relationship between the atomized individuals of technological civilization and the individualized warfare of the contemporary manhunt. Legions of psychologically disaffected and surplus populations fuel the Predator Empire. The fundamental question is not simply whether we are the masters or slaves of this robotic imperium but whether drones are creating a better, safer world for us to inhabit. The existential danger posed by the Predator Empire is not simply geopolitical blowback, then, but the dronification of the human condition. This names the mass production of anxious, hypersecured, and highly atomized individuals: soothed and yet ever distressed by the buzz of police robots swarming the skies.

1

The Long March to Human Enclosure

Technological Civilization

Elysium is a 2013 dystopian science-fiction film directed by Neill Blomkamp. A terra-formed space station, called Elysium, which is populated entirely by wealthy individuals, hovers over an impoverished and gutted earth. The economic and social distinction between the two spaces—one an artificial utopia shaped like a ring, the other a gritty slumscape—is the central antagonism of the plot. But what makes the movie memorable is not simply the scenes of immigrants attempting to reach Elysium nor the desperate images of a surplus population confined to a scorched planet earth, but the relationship human workers have with robots. The main character, a former car thief named Max Da Costa, works at an assembly line producing weaponry for the aerial drones and police robots that now scour the planet. The irony of the mechanical apartheid is here magnified: not only does a humanity rendered surplus by technology produce wealth for a distant, fortified, ruling class, but it also produces the very robots that enforce that segregation.

The Predator Empire emerges from the unprecedented ways in which human life is now mediated and secured by technical systems. It may not be the empire of our choosing, but it is the empire of our time: the Leviathan born of the electronic traffic that pulses across the globe and the drones that soar through the skies. Drones are becoming key technologies of state power for maintaining controlled societies and subjects. Accordingly, exploring the existential relationships between humans and nonhumans matters for how we understand power, geopolitics, and the Predator Empire. How are objects, weapons, infrastructures, and bodies mangled together in the contemporary technogeographies of state power? Enclosure, as both a historical act and a process of contemporary securitization, is an important starting point because it sets in motion many of the geographical and psychological contradictions of technological civilization. And underpinning the dyad of technological civilization and the Predator Empire is *imperial space*: the abstract, geometric order installed across the planet during the modern age and European colonialism. In this sense, U.S. drone warfare crystallizes longer historical trends that have sought to secure and contain humanity for millennia.

We now exist in the age of "electronic globalization" according to Peter Sloterdijk. The land- and sea-based terrestrial globalization, synonymous with the colonization of the planet after Christopher Columbus traveled to the "New World" in 1492, "ended with the installation of an electronic atmosphere and a satellite environment in the earth's orbit in the 1960s and 1970s."[1] Electronic globalization has ushered in a spatial revolution, one that has consistently eroded the importance of geographical distance for communication and the exercise of state power. Electronic surveillance, electronic policing, electronic warfare, and electronic assassination—these are the four horsemen of the Predator Empire that gallop across the earth. Since the twentieth century, railways and telegraphs, themselves productive of new space-times in the world, were overthrown by airplanes and radio signals. The geographies of human contact moved from the surface of the planet into its surrounding atmospheres. "For radio and light messages, the earth has virtually shrunk to a single point—it rotates, as a temporally compact orb in an electronic layer that surrounds it like a second atmosphere."[2]

This electronic atmosphere emanates from the technological civilization that now stretches over the planet like a cyborgian skin. With its enclosure of so much of the human species, technological civilization has inaugurated a series of profound existential shocks. Max Weber famously labeled the modern age's rationalization of society as the "disenchantment of the world." This existential condition, fed by an ascetic capitalism, empties the cosmos of its gods and replaces them with a pervasive scientific worldview. Humans are left unprotected on a planet bereft of existential shells. The heavens are replaced by an infinite and frosty sky. "Now that God's shimmering bubbles, the celestial domes, have burst, who could have the power to create prosthetic husks around those who have been exposed?"[3] Since religion no longer secures and controls the species as it once did, modern humanity "must now create its own satisfaction on artificial continents under artificial skies and domes."[4] In other words, technological civilization, which first killed the gods, must now protect its newly exposed denizens. It thus becomes a kind of immunitary system, a form of synthetic protection that aims "for an imitation of the now impossible, imaginary spheric security."[5] The profound existential security of modern living is contained by artificial domes and, as will become apparent, artificial drones.

To think through the causes and consequences of the Predator Empire thus requires us to investigate its relationship with technological civilization. Narrowly conceived, the Predator Empire guards the U.S. homeland against external threats (although even this formulation is complicated). Such a justification is used continually to feed the growth of an unprecedented national security state and a national security economy. Underpinning this rationale, of course, is a much broader depiction of an absolute sovereign defending civilization against an anarchic (and barbaric) outside. But this notion of the Predator Empire shielding civilization leaves unanswered what exactly civilization is. The use of "civilization" may seem problematic, and its use in international relations (IR) and beyond continues to generate heated debate. Despite this, *civilization* is a term continually invoked by state elites, particularly in the U.S.-led war on terror. A week after the September 11, 2001, terrorist attacks, for example, President Bush argued the war on terror was not a U.S. fight but a "civilization's fight."[6] As Mark Neocleous concludes, "The war on terror, as international ordering, is a form of police; civilisation's return writ large."[7] The Predator Empire's civilizing mission thus seeks to protect the U.S. homeland in particular and civilization more generally.

The term *Western civilization* has been used repeatedly throughout history. Today, it is still used to describe the congregation of governments, institutions, cultures, and economies that bind much of the global North. The idea of democratic freedom so central to this constellation is, however, determined increasingly by the supposed economic freedoms of capitalism. Indeed, U.S. geopolitical strategy often coincides with such geoeconomic calculations. "Securitization is invoked," argues John Morrissey, "via a universalist neoliberal logic to protect and safeguard the global economy."[8] Under this kind of understanding, defending civilization amounts to protecting the freedom of global capital. Perhaps, then, empire is a deterritorialized economic force. This is a form of what Michael Hardt and Antonio Negri call "capitalist sovereignty," in which the Leviathan dissolves into the immanent circulations of the global market.[9] "Empire is the political subject that effectively regulates these global exchanges, the sovereign power that governs the world."[10] In other words, if we understand Western civilization as a purely *economic* project, it limits the parameters for how we define empire.

As important as this economic understanding of empire is, then, and despite the continued rhetorical power of Western civilization, the active role of materiality in organizing and dominating humanity is left underappreciated. We must take stock of the carceral husks that entomb our neurotic fears and secure surplus populations. Humans are not just economic animals but spheric creatures that install immune systems and artificial civilizatory worlds. If, as Peer Schouten argues, "IR has been oblivious to the significance of material civilisation for the constitution and transformation of political order," then consider this book an attempt to address this shortcoming.[11]

Accordingly, the concept *technological civilization* best encapsulates our dense technical living and provides an important material foundation for thinking about state power in the modern age, while still driven by an understanding of economic inequality. A persistent trend in capitalism is to augment, replace, and devalue human labor with fixed capital. The mechanization of the factory in the nineteenth century fulfilled this function. Today, it is reflected in the ongoing robotization of economic activity within technological civilization. Capitalism in the twenty-first century is constituted by increasing levels of nonhuman capital, which is centralizing power in fewer and fewer human hands. Technological civilization is thus constituted by the artificial infrastructures that enclose ever more unequal populations. It brings to the surface a much deeper meditation on the specific role of state technology and other capital-intensive apparatus of control in mediating human subjectivity in a globalized world. Crucially, this understanding impacts how we conceptualize drone warfare, as well as political geography and international relations more generally. In other words, we need to reinsert the existential dimension to any conceptualizations of civilization and empire: the robotization of the economy, security, and subjectivity are all linked.

The modern age is, above all, defined by the embedding of technology in the most intimate and diffuse spaces of the planet. Technology has enabled social conglomerates to become ever more complex through nonhuman forms of mediation and control. "Human society," argues Graham Harman, "is in no way made solely or even principally of people, but requires fences, coins, uniforms, monuments, ships, flags, wedding rings, and highways in order to stabilize itself."[12] So while technological civilization inherits the great social war that began with the enclosure of the commons centuries ago, it must be understood also as a technological—rather than purely economic—constellation, even if the two forces are linked.

Of course, humans have always employed objects, tools, and machines in their social lives. As Bruno Latour has long argued, human society is nothing without nonhuman mediators; our anthropology is always already embedded in technology. Civilization, in short, has always been technical. So why use such a grand label as *technological civilization* in this book? Because we have reached a point in which state mediation is now *predominantly* nonhuman. This is why to speak of technological civilization is to outline a form of social and economic organization historically far more automated, bureaucratized, and robotized than previous human constellations.

Jacques Ellul, describing what he calls "technique," or the technical mediation of society, argues that "without exception in the course of history *technique belonged to a civilization* and was merely a single element among a host of nontechnical activities. Today *technique has taken over the whole of civilization.*"[13] This "extraordinary event," concludes Ellul, has created what he calls technical civilization. "*Technical civilization* means that our civilization is constructed . . . *for* technique (in that everything in this civilization must serve a technical end), and *is* exclusively technique (in that it excludes whatever is not technique or reduces it to technical form)."[14]

Agnes Heller lists several paradoxes associated with what she terms *technological civilization.*[15] On the one hand, it may relieve humanity of its many burdens, create vast wealth, and save a lucky few from punishing labor. On the other hand, it alienates humanity, pollutes the earth, and enslaves the human species to the drudgery of the machine. As she argues, the more we surround ourselves with a technical world, the more "appendage-like" our lives become. "This appendage-like, reactive, superficial life can be termed 'alienated' in the sense that the essence of the human individual becomes external to his or her existence, just as in the case of madness. Alienation is madness, although undetected, for if everyone is mad, madness goes on unnoticed. Technological civilization empties out life through alienation."[16] Technological civilization, however wondrous, rests upon this kind of undetected madness.

Technological civilization must therefore continually pacify its own populations to secure its continued survival. To be civilized is to have everyday life mediated—sometimes brutally—by technical infrastructures. This is a process of mass *individuation* in which individuals are physically and psychologically reproduced inside their civilizatory spheres. The paradox, of course, when dealing with any notion of civilization is that the actual process of civilizing rests on its own kind of violence. Sigmund Freud was an early observer of the psychological contradictions thrown up by living in a civilization that is fundamentally oppressive.[17] The passions of our apish ancestry are sublimated in the name of a higher authority—whether god, monarch, or nation. As he wrote in 1929, "Our so-called civilization itself is to blame for a great part of our misery."[18] As more and more jobs are replaced by nonhuman capital, the expelled find themselves policed, occupied, and watched by an equally robotic security armada. And in between these technics swells a profound discontent. In this sense, a robotizing technological civilization renders vast swathes of humans materially and psychologically insecure.

Herein lies the essential role of the Predator Empire: as well as eliminating more obvious external threats—such as so-called Islamic State terrorists in Syria—it must also quash its internal threats, which range from the so-called war on drugs along the U.S.–Mexico border to the endemic social war played out across America's streets. The crises the Predator Empire addresses are simultaneously inside and outside the U.S. nation-state and both overly militaristic and implicitly civil in nature. "In contrast to imperialism," Hardt and Negri write, "Empire establishes no territorial center of power and does not rely on fixed boundaries or barriers. It is a *decentered* and *deterritorializing* apparatus of rule that progressively incorporates the entire global realm within its open, expanding frontiers."[19] Anything that threatens the enclosed civilizatory order can be targeted. The U.S. war on terror and the drone warfare it has spawned are both symptoms of this deeper existential malady. By now, two important concepts are in motion: the Predator Empire and technological civilization. Both are interlocking, and understanding them together deepens our understanding of drone warfare as an existential condition.

American Empire?

The Predator Empire complicates and advances popular and historical understandings of American empire by taking the nonhuman—

whether drones, computers, or algorithms—seriously in the execution of contemporary imperialism. Empires are traditionally understood as the spatial expression of imperialism: the control of one state over another state. This domination can take many shapes, from political influence over client states to more directly administered colonies. This split between a center of power, or metropole—from the Greek *metropolis* (mother city)—and a periphery has traditionally been a defining characteristic of empire. It's a schism that conveys a sense of a geographical inequality. The idea of empire today—American or otherwise—is extremely contested. As Simon Dalby argues, "If empire is to be of analytical use, it seems that none of the versions of it in circulation are quite adequate to grapple with nuances of the present."[20]

Certainly, the classic division between a metropole and a periphery has given way to more hybrid forms of sovereignty and control. John Agnew, for example, argues hegemony and empire are analytically distinct.[21] The former is based on soft power generated by common rules, values, and institutions. Empire is, however, "anchored by military coercion" and strongly territorial. This military coercion, in turn, has been accompanied always by a geographical imagination. Typically, this geographical imagination has pivoted around the idea of European civilization clashing with savages on its frontiers. Edward Said writes empire would not exist "without important philosophical and imaginative processes at work in the production as well as the acquisition, subordination and settlement of space."[22] Empire means, in short, a lot of different things to a lot of different people. The United States embodies these kinds of contradictions: it was baptized in the blood of Native American genocide but also in rebellion against British rule.

The nascent U.S. republican model broke with eighteenth-century Europe and was designed to prevent the establishment of a dictatorship. But it offered little guidance for restricting activities beyond U.S. borders. After the "closing of the frontier" at the end of the nineteenth century, the impulse to push westward gave way to a broader project of maritime policing. Both of these ideas contributed to the 1898 Spanish–American War, after which a victorious United States acquired the Spanish territories of Cuba, Puerto Rico, Guam, and the Philippines. This network of newly inherited islands and a growing navy allowed the U.S. military to take advantage of an enormous watery territory, just as the Portuguese and British empires had done. The suppression of political movements in Latin America, so common in the Cold War, has its roots in this oceanic expansion.[23] Indeed, during President Woodrow Wilson's tenure (1913–21) the Monroe Doctrine of 1823—which asserted the United States was the guardian against European intervention in the New World—was reworked as an idealistic vision for spreading democracy around the world.

Wilson strongly believed in the exceptionalism inherent within the United States. This belief was shepherded through World War II, after which the United States emerged as a global superpower. Germany's defeat represented something of a crossroads: the United States could return to a nominal isolationism or consolidate its newly won expansionism. Under President Harry Truman any chance of brokering peace with Joseph Stalin soon stalled (peace may have been a possibility if Henry Wallace had succeeded Roosevelt). Instead, a grim parade of proxy wars and conflicts—from Vietnam to Latin America—soon swallowed the planet as the United States installed its Cold War military enclosures into the soil of the earth.

Modern neoconservatism reworked the idealism of Wilson into a preventative doctrine of military intervention. By invading Iraq the Bush administration adhered not only to a "brazen imperial strategy" but "a new ideological commitment to empire," according to Agnew.[24] Consider the extraordinary powers granted to the U.S. president under the September 2001 Authorization for the Use of Military Force (often abbreviated to AUMF). This legislation granted the president the discretion to "use all necessary and appropriate force" against any nation or individual "he determine[d]" was connected to the terrorist attacks of September 11, 2001. Nowhere does the constitutional model of U.S. government seem more under pressure than with the rise of such a monarchical executive, a topic explored in more detail in chapter 3.

Perhaps, the Bush years should not distract us from what Andrew Bacevich calls the underlying Washington Rules: "the national security consensus to which every president since 1945 has subscribed."[25] This consensus has played its part in U.S. military interventions across the world and has legitimated the installation of brutal dictators to further American interests in Latin America and beyond. As Chalmers Johnson writes, "The unintended result of this record of militarism is the contemporary Leviathan that dominates Washington . . . disgracing the nation by allowing our young men and women to torture prisoners picked up on various battlefields or even snatched from city streets in allied countries."[26]

Many sympathetic commentators characterize the United States as an informal, or reluctant, empire because it lacks the colonial possessions of former world powers. Certainly, for much of the nineteenth and twentieth centuries, the United States sought to establish markets rather than colonies.[27] Other commentators find the idea of an American empire implausible in an era of globalized capital. As Hardt and Negri write, "First of all, the coming Empire is not American and the US is not the center."[28] But perhaps this globalization narrative underplays the geopolitical clout the U.S. military has continued to hold for decades. As Matthew Sparke argues, "Many writers . . . gloss over the asymmetries and uneven development patterns associated with American dominance."[29] Collapsing the two narratives together, Neil Smith argues, "Today we are living through a third moment of US global expansion, best encapsulated in the language and ambition of a US-led globalization."[30]

Indeed, despite appearances, ever since the 1898 Spanish–American War, the United States has ruled over a vast web of island territories, which includes islands across Puerto Rico, Guam, American Samoa, Micronesia, the Marshall Islands, the Commonwealth of the Northern Marianas, Palau, the U.S. Virgin Islands, and many more. The residents of these archipelagos are treated typically as second-class U.S. citizens and have no voting rights or representation in the U.S. Congress. This status stems from the Insular Cases at the turn of the twentieth century, which ruled that within unincorporated territories, inhabitants were neither aliens nor citizens. Ruth Oldenziel calls this archipelago of islands a "networked empire." The U.S. military extended its control of the ocean during the twentieth century, creating "a global system of international relations in which islands, peninsulas, and littoral spaces played a key geopolitical role."[31]

This allowed the U.S. military to circumvent claims of imperialism while benefiting from an expansive territorial network. Under a series of partnership deals in the twentieth century, the United States leased the British Empire's ocean network of outputs, bases, and ports. The British Navy had first pioneered a technopolitical form of ocean power by installing a system of underwater cables and coaling stations. This is one instance of technopolitics that takes seriously the ways in which state power becomes materialized in technological systems. As Gabrielle Hecht argues, "The material properties of technologies shaped the exercise of political power in the second half of the twentieth century."[32] By the mid-1950s the Pentagon articulated the Strategic Islands Concept and continued to lease littoral spaces across the planet and fill them with advanced military technology. These "emptied" archipelagos—the indigenous populations were usually "resettled"—were again essential for the 1968 Satellite Triangulation Program. As Oldenziel concludes, "Anchored in 'empty' islands, the reach, power, and prestige of large technical systems—from the first telegraph communications to current outer space systems—have come to replace territorialized empire as an indicator of geopolitical power."[33]

Similarly, for Johnson, American empire is founded on its network of military bases across the planet: "This vast network of American bases on every continent except Antarctica actually constitutes a new form of empire—an empire of bases." As he adds, "Without grasping the dimensions of this globe-girdling Baseworld, one can't begin to understand the size and nature of our imperial aspirations."[34] Although there was a reduction in U.S. military bases after the occupation of Iraq and Afghanistan, this has been accompanied by a growth in smaller, leaner skeletal bases in remote areas of the globe. These are sometimes referred to as "lily pads," since they allow the military to hop across the planet and maintain "global dominance by doing far more with less," according to David Vine.[35] As Sasha Davis puts it, the lily pad strategy enables the United States to strike anywhere, anytime, "without any need for consultation with anyone."[36] For Tom Engelhardt these lily pad bases, which exist across northern Africa all the way to the Chinese border, are part of a military strategy "meant to encircle and nail down control of this vast set of interlocking regions—the thought being that, if the occasion arises, the American frogs can leap agilely from one prepositioned pad to another, knocking off the 'flies' as they go."[37] The dronification of state violence requires us to rethink how we understand empire, foregrounding the geopolitical power of nonhumans.

The new face of the U.S. military's empire has far fewer human faces. After the brutal counterinsurgencies of Iraq and Afghanistan, Washington's war managers scrambled to find a new strategy. Their response was to do more with less, reshuffling the geography of human soldiers, robots, and drones—shifting vulnerable U.S. bodies from the frontline to the back office and therefore radically redistributing the precarity of life. In other words, American empire was transforming—and had been for decades—from a labor-intensive to a machine-intensive system of dominance: the Predator Empire. Drones, for example, enable a lighter military footprint but can still produce a widespread surveillance of the planet. Indeed, they have quickly become a preferred method for carrying out foreign attacks. In the conflict against the Islamic State, for example, the U.S. military has pursued mixed air and drone strikes across Iraq and Syria, which builds upon the CIA's targeted killing program in Pakistan's tribal areas. And so, even if the U.S. military withdraws its human troops from the Greater Middle East, it will leave in place the architecture for a forever drone war.

Toward a More-than-Human Geopolitics

Building on the introduction's reworking of Hobbes's Leviathan, we must further explore how and why objects and technical systems possess the power to intervene and reshape international relations. This connects to deeper philosophical arguments about materiality. Hannah Arendt, for example, writes, "The objectivity of the world—its object- or thing-character—and the human condition supplement each other; because human existence is conditioned existence, it would be impossible without things."[38] The radicality of this statement is to define human existence as inseparable from the "thing-character" of the world. Accordingly, we need to construct a more-than-human geopolitics centered on the existential power of things.

The idea of a more-than-human geopolitics is used to signal that the exercise of global power is both enabled and conditioned by technological, or more-than-human, systems. This complicates the notion humans are the sole means by which geopolitical power is performed and steers our analysis away from the realist and instrumental views of drone warfare. In doing so, it does not evacuate the importance of the human but places them within wider assemblages of nonhuman power. A more-than-human geopolitics therefore necessitates an understanding of foreign policy that isn't overly influenced by physical geography (as was the case with the classical geopolitics pioneered by Halford Mackinder) but includes machines, computers, weapons—indeed, the armada of objects that sail across the electric seas of technological civilization. Perhaps most important, a more-than-human geopolitics is grounded in the belief that technology is always an existential force.

Bruno Latour's actor-network theory (ANT), while never outlining a sustained engagement with violence or warfare, is a crucial entry point into a more-than-human geopolitics. The key insight he offers is an understanding of how nonhuman entities are embedded within the political fabric as engines of social securitization. As he writes, "In addition to the throng of little people summed up in the crowned head of the Leviathan, there are objects everywhere."[39]

This sensitivity to nonhumans is not usually figured in how geopolitics and international relations are typically conceptualized. Erika Cudworth and Stephen Hobden argue, "International Relations remains dominated by humanocentric approaches."[40] And Claudia Aradau adds, "Although analyses of security and risk have incorporated discussions of technologies and institutions, non-human objects have been relegated outside the realm of securitization."[41] Of course, whether gunpowder or the atom bomb, technology has frequently been a topic in international relations (IR) and political geography. Usually, however, technology is viewed as merely embodying social relations. This is an instrumental view of the nonhuman that holds technology expresses human intentionality. To give an example, according to Mike Bourne, "Weapons are generally portrayed as mere artefacts: the passive tools of states' relations in anarchy."[42]

Yet it doesn't take long to realize we are surrounded by a multitude of nonhuman objects that secure the human condition and complicate the lines between human and nonhuman agency. Indeed, technological civilization would be impossible without the power of things to intervene in our lives. Crucially, the passage of time has seen state violence, surveillance, and policing augmented with technical mediators. With phenomena as diverse as drone warfare, climate change, and the 2014 Ebola outbreak, a recognition is growing that human security and geopolitics are very much interdependent with the nonhuman. The great division in philosophy and IR, alike, between the human and the nonhuman, or between society and nature, is coming undone.

The material turn in philosophy and the social sciences has attempted to redistribute agency to nonhuman actors and forces. There are lots of isms in these intellectual schools of thought: posthumanism, antihumanism, new materialism, and speculative realism. These are accompanied by complexity theory, object-oriented philosophy, technopolitics, and the old stalwart ANT. While all of these theoretical frameworks are different, they usually challenge the anthropocentric view that the human is the *only* way to make sense of the world. "In doing so," argue Nick Srnicek et al., "the materialist position is led to ask critical questions about rational actors, agency in a physical world, the role of affect in decision-making, the biopolitical shaping of bodies, the perils and promises of material technology, the resurgence of historical materialism, and the looming environmental catastrophe."[43]

As such, whether with concepts like *technology, infrastructure, objects, actors,* or *things,* we are dealing with nonhuman forces that continually remake and disrupt the geopolitical landscape. "The struggle between actors," argues Harman, "must include nonhuman ones as well: not just forts and weapons, but atoms, machines, rainbows, buses, and tar. We no longer accept a two-world physics of superlunary and sublunary realms; by the same token, we should not accept a two-world politics in which human power struggles are treated differently from the duels between humans and nonhumans."[44]

Srnicek uses the concept *momentum* to stake a middle ground between a view of technology that is either instrumental (i.e., socially constructivist) or deterministic (the nonhuman is entirely autonomous): "Momentum stems in part from the fact that technologies have a logistical footprint: any given technology implies an entire system of production, distribution channels, technical experts, and subsidiary technologies. A technology, in other words, always already necessitates a larger sociotechnical assemblage, and as a result a set of shifts that emerge from adopting it."[45] A realignment toward a more-than-human geopolitics thus requires us to redistribute agency away from purely human relations to the technologies, or technogeographies, that mediate humanity across the world and generate sociopolitical momentum.

Furthermore, the international system must be seen as a dynamic—rather than static—place in which nonhuman entities continually transform the conditions for violence, communication, and warfare. In this way technical mediation is an active force in international relations. Technology is agential, since it creates tendencies, trajectories, and conditions that impact hybrid sociotechnical assemblages. This could be defined as *nonlinear technological determinism*. While technology certainly *does* cause social transformations, the direction of these cannot always be predicted in advance. In other words, while technology is crafted by humans, it obtains a degree of technological autonomy by locking in a particular momentum. And since complex systems are nonlinear, "the implication of this is that very small actions by agents can potentially have rather far-reaching effects."[46]

Material infrastructures are central to the maintenance of international relations of power. "Objects, people and states," argues Bourne, "do not merely and passively express the will of some deeper and prior structure or interest; and ideas, norms and values cannot act without materialisation."[47] The crushing inequalities and violences tattooed into the flesh of the world depend not simply on human culture, imagination, and ingenuity but on its past and present materializations. This means the nonhuman is very much enrolled in the world's geopolitical organization. As Karen Barad writes, "To restrict power's productivity to the limited domain of the 'social,' for example, or to figure matter as merely an end product rather than an active factor in further materializations, is to cheat matter out of the fullness of its capacity."[48] Or as Aradau concludes, "In this light, securitization needs to be understood as a process of materialization that enacts a reconfiguration of the world in ways in which differences come to matter."[49] In the next section I examine just why this *matters*.

Why Does the More-than-Human Matter?

The Predator Empire cannot obtain domination without first materializing its apparatuses into the planet—across land, sea, outer space, and even cyberspace. The matter matters, enacting a reconfiguration of the world. In this sense it is extremely important to see the historical and the technological as simultaneous forces. "Matter is always already an ongoing historicity."[50] It is with this "ongoing historicity" in mind that U.S. drone warfare must be understood. What does drone warfare enable the state to do that it couldn't do before? What geographies does it inscribe into the landscape? While it's easy to dismiss drones as tools under the control of human beings, it is important to see them as geopolitical actors in their own right.

The use of drones was cited as a reason why the 2011 U.S. military intervention in Libya circumvented the 1973 congressional War Powers Resolution (a law that restricts the president's ability to unilaterally go to war). The logic was as follows: because there were no troops on the ground, it wasn't a war. Former legal adviser to the Department of State Harold Koh insisted that "the limited nature of this particular mission is not the kind of 'hostilities' envisioned by the War Powers Resolution."[51] A precedent with the Libyan air strikes was set in motion. By removing U.S. bodies from the battlefield, politicians "no longer treat the previously weighty matters of war and peace the same way," according to Peter Singer. As such, "we possess a technology that removes the last political barriers to war."[52] Drones, by enabling a form of remote war, possess the unique potential to undermine the enshrined checks and balances of democracy.

Technology, therefore, endlessly mediates and dislocates social norms. A substantial part of our ethical universe rests upon the shifting tectonics of technological apparatuses. Far from floating above the clouds, ethics is very much a labyrinth of technological artifacts populated by nonhuman forces that intervene upon the thoughts, opinions, and cognitions of society. Technology is, in other words, a mode of existence or a transcendental condition that underpins violence, law, and ethics. Crucially, this means drone technology is continually translated into a *legal* capacity. International law has been—and continues to be—challenged and reengineered by unmanned conflict. As David Kennedy argues, "Law has an effect—is law—when it *persuades* an audience with political clout that something someone else did, or plans to do, is or is not legitimate."[53] Law is always a process, an action, and therefore part of a wider sociotechnological momentum.

Drone technology has been central to the legal justifications made by the White House. John Brennan, director of the CIA, has insisted targeted strikes are ethical: "With the unprecedented ability of remotely piloted aircraft to precisely target a military objective while minimizing collateral damage, one could argue that never before has there been a weapon that allows us to distinguish more effectively between an al-Qa'ida terrorist and innocent civilians."[54] The ability to target with precision has led to the belief drones are ethically superior to unguided bombs. Technological precision, in other words, is implied frequently as compliance with international law.[55] Yet the targeting capability of a drone tells us nothing about whether a target is actually a combatant or whether there is any military advantage in carrying out the strike. A danger exists that ethics is being overrun by technics.

Drone warfare is an important example of the philosophical notion of a means transforming the ends of an action. Concepts like *surveillance* and rhetorical devices like *threats to national security* are retroactively altered by the means used to realize them. As an example, the Predator drone allows the U.S. military to surveil populations thousands of miles away and discover, in rugged mountains and hostile deserts, threats to national security. This means the epistemological problem of what, where, or who is dangerous is inseparable from the tools used to pose that problem. Surveillance, in other words, does not simply discover danger but simultaneously reproduces it. By delegating the art of killing and surveillance to nonhuman means, humanity is slowly carried along by the momentum of its creations. The diffuse battlespace of the war on terror, one defined by a global manhunt and an amorphous temporality, is enabled by the very technologies deployed by the U.S. military.

Moreover, since drones are occupying the position of the frontline U.S. solider, the ideals of heroism, courage, and sacrifice are being transformed. As we are repeatedly reminded, the inverse is now true: sparing U.S. soldiers from frontline action is now virtuous. The preservation of the American way of life in a dronified war on terror has meant assassination has taken center stage in U.S. national security—exemplifying what Grégoire Chamayou calls "necroethics." As he argues, the U.S. drone pilot has been reduced to the figure of the executioner, since the very possibility of violence against her or him is precluded: "The paradox is that the drone, so highly praised for its great ability to make out the difference between combatants and noncombatants, in practice abolishes the very condition for that differentiation, namely combat."[56] The drone matters because its remote killing structurally precludes reciprocity as such, which in turn reengineers the modality of state violence to a naked form of killing with impunity.

Drone warfare is not the first more-than-human conflict, of course. In fact, war has always been a melee of flesh and things.[57] But drone warfare does intensify the complex relationships between humans and machines. Caroline Holmqvist argues for a "critical materialist reading" of drone warfare based on a phenomenology in which the drone "acts" by transforming the way violence is represented and conducted.[58] By focusing on how drones mediate military vision and violence, we begin to appreciate how the human is constantly undone by war technology. The drone is, in other words, *performative*: it creates a disembodied, highly mediated cosmic view that converts human lives into crosshairs and digital patterns. It is in this sense, as Majed Akhter and I have written, that "the drone performs the military logic of a 'war without the war' to its extreme, which is to say, a war without bodies, a war of machines, and a war of discrete and surgical strikes from the sky."[59]

Why Does the Geo Matter?

It is important to conceptualize *geopolitics*—rather than simply *politics* or *international relations*—when analyzing drone warfare. Typically, whenever IR discusses objects, things, or their networks, the spatiality of their effects tends to be ignored in favor of analyzing the role of discourse, treaties, or the personality of state leaders. Power, however, must always be considered as spatial: it is immanent to the infrastructures materialized within a more-than-human landscape. Geopolitics has long foregrounded the geographical nature of state power, even if it has tended to focus narrowly on physical features of the planet (mountains, oceans, rivers) at the expense of the wider realm of nonhuman forces. A keen attention to a more-than-human geopolitics of drone warfare thus unlocks the atmosphere of its effects and helps explain the crushing geographical asymmetries that nonhuman actors can maintain across the planet. Drones, when understood through a more-than-human lens, become more than a tool of rational state actors waging a clearly defined war. Instead, they transmute into a kind of spatial—or *spherical*—power that spreads across technological civilization, endlessly reprogramming the environments of human coexistence.

State power is thus inherently spatial. Michel Foucault first labeled the state's management of life as "biopower." With this term he refers to a set of government mechanisms through which the biological features of the human species become targets of political power. Unlike overtly violent forms of sovereign punishment, biopower manages our physical health, mental well-being, and social conduct. If biopower targets the human as a biological species, then it necessarily administers the surrounding lifeworld of the organism, regulating the atmosphere of its existential spheres. Biopolitics' last domain, as Foucault writes, is "control over relations between the human race, or human beings insofar as they are a species, insofar as they are living beings, and their environment, the milieu in which they live."[60] For Sloterdijk this idea can be boiled down to a single axiom: "biopolitics begins as enclosure-building."[61] Indeed, as the Predator Empire fossilizes in the bedrock of the planet, it entrenches a technological momentum, or a machinic enclosure, beyond the control of any single person, army, or state. Enclosure-building, or the securitization of the atmospheres of humanity, is thus a central concern of a more-than-human geopolitics.

We pass through a variety of spheres throughout our lives. At birth we exit from our primordial sphere—the womb—and are thrown into a universe indifferent to our existence. As discussed at the beginning of this chapter, the Western Enlightenment tore down the gods from the heavens, further stripping humanity of its protective spheres. Exposed, vulnerable, and alone, humans shielded themselves with a whole manner of prosthetic husks. Humans became existential builders "continually working in their accommodation in imaginary, sonorous, semiotic, ritual and technical shells."[62] A profound insecurity of life on the outside is tranquilized by living in artificial enclosures. This is the central, *dome-estic* truth of technological civilization. "The body of humanity," writes Sloterdijk, "seeks to create a new immune constitution in an electronic medial skin."[63]

Modern humanity thus "attempts to balance out its shellessness in space, following the shattering of the celestial domes, through an artificial civilizatory world."[64] *Who* we are is therefore really a question of *where* we are. And herein lies the key point: the space of humanity is never simply human. "To live together in the world," writes Arendt, "means essentially that a world of things is between those who have it in common, as a table is located between those who sit around it; the world like every in-between, relates and separates men at the same time."[65] The human condition is torn open by things and patched together by things. An armada of technologies encloses the atmospheres of human existence. A regime of machinic discipline—ruled by nobody in particular—organizes life inside technological civilization.

As I make clear in the following section, technological civilization's war is always already a war of enclosure—a neurotic impulse to bring humanity *inside* its securitized spheres. In order to trace the ongoing historicity of an enclosed civilizatory materialism, the next section presents a brief history of the practice of enclosure. The wager this book takes is that the military drone is not simply a weapon but a geopolitical actor that seeks to enclose the world by producing overlapping, electromagnetic, civilizatory domes. This momentum has been building for some time, and its totalitarian impulse lies not in a single coup d'état but in its insidious spread in the intimate spaces of everyday life. As a more-than-human geopolitics makes clear, the wings of the Predator Empire spread far and wide, soaring high into the skies, before diving into the deepest fibers of human being.

A History of Human Enclosure

Exploring enclosure establishes the ground zero, or the fundamental existential condition, upon which drone warfare is built. *Enclosure* conveys multiple and overlapping meanings, all of which this book plays upon. Typically, *enclosure* is defined as a space surrounded by a barrier, a kind of inside in some way protected from an outside. Whether a thorny medieval hedge surrounding a field, a rotten wooden fence circling a house, or a billion-dollar surveillance wall in the Arizona desert, each of these apparatuses produces an inside and an outside, impeding the free movement of people. Accordingly, before U.S. drone warfare is considered in isolation, it must be situated within a longer historicoexistential trajectory. It is dangerous to separate drone warfare from the illnesses of our technological civilization. To repeat Sloterdijk, "Biopolitics begins as enclosure-building."[66] And so, even if the drone's geographies are new and its effects far reaching, it is a technology embedded within the long march of human enclosure, of imprisoning life inside nonhuman apparatuses.

Remaking the world into a system of secured insides and dangerous outsides required a protracted, violent deworlding of millions of lives, using a range of apparatuses. Apparatuses materialize the potential forms of life into distinct enclosures: they are the "making of spacetime itself."[67] If human life on earth was once open and rich with possibility, it has since become enclosed within apparatuses of growing sophistication and scale. As Giorgio Agamben writes, "Today there is not even a single instant in which the life of individuals is not modeled, contaminated, or controlled by some apparatus."[68] These range from prisons and asylums to computer databases and CCTV cameras. An important site of historical enclosure is the English enclosure movement, which privatized commonly used tracts of land at the dawn of the modern age and through the Industrial Revolution. "The 'English enclosure movement,'" Peter Linebaugh argues, "has belonged to that series of concrete universals—like the slave trade, the witch burnings, the Irish famine, or the genocide of Native Americans—that has defined the crime of modernism."[69]

The history of English enclosure is contested and subject to repeated revision. Even terms like *commons, commoner,* and *peasant* are not settled. Partly, this is a result of fragmentary records, but also, the history of enclosure was largely written by those who engineered its success. But whatever economic rationale was given for or against enclosure, it was never a purely economic project: it was always a form of social war, a project of existential engineering.

Long before enclosure, feudalism was the dominant pattern of land ownership across much of medieval Europe. At the foundation of the feudal system was the lord of the manor, who ruled over serfs and peasants laboring in the surrounding fields. Manors often had common land the lord did not use directly. During this period the open-field system of farming was popular, with peasants working arable strips of land together. While this land use was hardly equitable, communities enjoyed common rights of tenure to graze livestock, grow crops, cut trees, and draw water from the ground. This was even recognized and protected by the English courts. For example, the 1217 Charter of the Forest—a companion document to the 1215 Magna Carta—protected common access to forestland. Villagers were able, in short, to carve out an autonomous existence for themselves with the customary and common rights they enjoyed.

The first assault on the commons began during the thirteenth century. Landlords began to convert common land into sheep pastures for their commercial use after a boom in the price of wool. "The motives of enclosers varied," writes Stuart Hodkinson, "but the desire to privatize the riches of the soil was omnipresent."[70] This first wave of enclosure had parliamentary backing, starting with the Statute of Merton in 1235. This piece of legislation paved the way for the subsequent dispossession of commoners. The Commons Act of 1285, for example, cemented the right to enclose land for individual use. A number of uprisings ensued, including the revolutionary 1381 Peasants' Revolt, which was at least partially attributable to enclosure and the dire conditions commoners faced in the wake of the merciless Black Death. The woodlands, pastures, wastelands, and farmlands that oxygenated the rural way of life were being slowly strangled.

Throughout the fifteenth, sixteenth, and seventeenth centuries, an enormous amount of land was subsequently enclosed. Agricultural society was becoming picked apart and atomized. The most definitive wave of land enclosures occurred between the seventeenth and nineteenth centuries with parliamentary enclosure. Whereas previous bouts of enclosure were designed to convert arable land into sheep pasture, the impetus now was to convert open fields, pastures, and wastelands into "productive" arable land, often for the "national good." The sheer scale of land change in this period is unsurprising: for generations landowners sat as peers in the House of Lords, effectively directing legislation coming out of Westminster. As Simon Fairlie writes, "The 'democracy' of late eighteenth and early nineteenth century English parliament . . . proved itself to be less answerable to the needs of the common man than the dictatorships of the Tudors and Stuarts."[71]

Between 1604 and 1914, over 5,200 enclosure bills were enacted in Parliament, relating to an area one-fifth the size of England, or some 6.8 million acres.[72] In 1873 *The Return of Owners of Land* was published. This book was the first profile of land distribution since the Norman Domesday Book of 1086. The controversial tome revealed an aristocratic elite now utterly monopolized the agricultural land in the country. Britain was largely enclosed: commoners were brought inside a growing technological civilization, stripped of their land and the rights attached to it. In order to survive, once-autonomous individuals now had to sell their labor and buy back the products and sustenance they once created themselves. People who were no longer able to subsist from the land were forced to sell their labor to survive. Karl Marx famously argued this historical act of expropriation was "written in the annals of mankind in letters of blood and fire."[73] The agricultural revolution was, in short, always an existential revolution.

But it wasn't a revolution installed without contradiction or struggle. Ruling elites grew concerned about the growing army of homeless and vagrant people. Enclosers realized they needed to work upon the newly enclosed terrain to mark and defend it from the exiled masses. Hedges were planted to form barriers and separate the land into distinct enclosures. Although they had long been used as windbreaks for crops and shelter for animals, the hedge took on a new, insidious character—which is why it was so often a target of resistance. As Nicholas Blomley writes, "The thorn hedge sought to protect private property from the bodies of the poor and became an instrument of class discipline, put to work in socially directive ways."[74] The hedge, in his words, "was a powerful machine of enclosure" whose hawthorn spikes formed "organic barbed wire."[75] So while enclosure relied heavily on the legal and cartographic powers emerging from Parliament, it was above all a material form of domination—a geopower—that shattered the ancient links between human and earth.

Resistance was common throughout the enclosure movement. Digger groups sprouted up across England, regularly destroying the hedges and ditches that policed the enclosures. For example, the True Levellers were a Digger group led by Gerrard Winstanley. The poor had, he insisted, the universal right to freely cultivate the land. Diggers set about reclaiming enclosed land, starting with St. George's Hill in Surrey. Winstanley wrote the first Digger Manifesto in 1649. An early example of socialist thought, the document insisted the earth was a common treasury and no branch of mankind should rule above any other. Winstanley warned, "That England is not a Free People, till the Poor that have no Land, have a free allowance to dig and labour the Commons, and so live as Comfortably as the Landlords that live in their Inclosures."[76] Although the Digger movement was relatively narrow in focus, it epitomized a backlash against the hollowing of the commons, the destruction of communities, and the terror born of abstract territory.

Newly "liberated" working families were drawn to the smoky labyrinths of the Victorian city or else consigned to starvation. The impoverished were now utterly dependent upon the very system that uprooted them. As Hodkinson argues, "Enclosure's role was essential as the physical-legal process that smashed the protective shield of common right that previously protected the peasantry from total wage dependence."[77] This orthodox understanding of what Marx calls "primitive accumulation" is now understood not only as the historical precondition of capitalism but as an ongoing act of what David Harvey calls "accumulation by dispossession."[78] As Linebaugh argues, expropriation and exploitation are not successive stages of capitalism but intensifying and interweaving processes that exist today in violent combinations: "The war machine and the machines of war, that military–industrial complex, arise from attempts to destroy the world's commons."[79]

Indeed, writers in the 1990s began to discuss the existence of "new enclosures." These are the ongoing spaces of class struggle where public resources across the planet are targeted by privatization: "Enclosure is one process that unifies proletarians throughout capital's history, for despite our differences we have entered capitalism through the same door: the loss of our land and of the rights attached to it."[80] Enclosure today works across multiple spaces: from the landscapes that nurture life to the DNA that codes life. As Alex Jeffrey et al. insist, "Whether in the form of resource wars, the often violent seizure of public lands for private capital, in bio-piracy, the destruction of the global environmental commons, or the revanchist onslaught of public services across the global North, capitalism is dependent on the division, conversion and demolition of various forms of public life."[81]

In other words, enclosure hasn't gone away, whether understood as primitive accumulation, accumulation by dispossession, or mass alienation. The common existential heritage we share is this loss of world, and like the commoners of England, we have become reliant on the very system that exiles us. Enclosure is, in short, a legal act, a secured space, and a disciplinary force. Each demands the other, and together they underpin the unstable technological civilization we call home. As becomes clearer in the following section, the practice of enclosure was exported across the globe as English enclosure became a template for colonial containment.

Exporting Enclosure to the World System

Enclosure spread throughout the world, colonizing distant horizons and infecting people with the same malady that rocked the fields of England. Communal forms of social organization were seen as backward and in need of economic salvation. "Security of property," argues E. P. Thompson, "is complete only when commons come to an end."[82] British colonists were key vessels in this system of expropriation. "The discovery of gold and silver in America," writes Marx, "the extirpation, enslavement and entombment in mines of the indigenous population of that continent, the beginnings of the conquest and plunder of India, and the conversion of Africa into a preserve for the commercial hunting of blackskins, are all things which characterize the dawn of the era of capitalist production."[83] Under this colonial imagination, the English peasant and the "wild" native were both resources that needed to be economically improved. This logic fueled the genocide of Native Americans, the plunder of Africa, and the installation of British law across the South Pacific.

On the one hand, European colonialism was a straightforward case of land theft. As Achille Mbembe explains, "*Colonial occupation* itself was a matter of seizing, delimiting, and asserting control over a physical geographical area—of writing on the ground a new set of social and spatial relations. The writing of new spatial relations (territorialization) was, ultimately, tantamount to the production of boundaries and hierarchies, zones and enclaves; the subversion of existing property arrangements."[84] On the other hand, space itself was fundamentally reimagined. In New Zealand, for example, the Native Land Act of 1865 transformed customary rights into individual property titles. Māori land was deworlded by a foreign register. The project of European colonialism was realized most violently when the planet's rich tapestry of lifeworlds—spheres of unique cultural time-spaces—was conquered by an all-encompassing, undifferentiated globe of economic exchange. As Sloterdijk explains, "Mastering space means eliminating its separating function and putting it to work exclusively as a conductor."[85] A global ground zero—a universal conductor—had to be produced for a global capitalism.

As the longitude and latitude of this new world order spread, the loot floated back to Europe. All things on the planet could now be understood through a universal measure—the currency of exclusive property. The *vacuum domicilum* was an English legal doctrine that enabled lands not "occupied" to be lawfully appropriated: a perfectly legal shock doctrine. As Edward J. Thompson remarks, "The same era that saw the English peasant expropriated from his common lands saw the Bengal peasant made a parasite in his own country."[86] As with the commons, displaced people in the colonies had to purchase back the products they once crafted with their own hands. Colonialism across distant shores mirrored and magnified the clearances taking place at home. "This taking, this expropriating the common," writes Linebaugh, "is a process of war, foreign and domestic."[87]

In the so-called Scramble for Africa, the European colonial powers inscribed into international law a civilizatory process of enclosure. The "wasteful" commons were written across a "barbaric" African continent in need of civilization. "At the end of the nineteenth century," argues Mike Davis, "the forcible incorporation into the world market of the great subsistence peasantries of Asia and Africa entailed the famine deaths of millions and the uprooting of tens of millions more from traditional tenure."[88] Economic improvement was the weapon of the colonizers, and it was partly legitimized in international law.

"This is why the central theme that so animated the early law of nations, the question of *just war,* is shot through with the categories of the war on the commons and the language of enclosures."[89]

In North America the idea of the open frontier was a crucial legitimizing trick for enclosers. "Just as the land must be cleared of trees and rocks in order to farm it," write Hardt and Negri, "so too the terrain must be cleared of the native inhabitants."[90] Colonialism of the Americas fused together a distinct European legacy of sovereignty as enclosure, divine church authority, and the doctrine of discovery. America was seen as a tabula rasa, a *terra nullius,* and the idea of civilization provided a framework for exporting the gift of enclosure and subjugating non-European populations. Civil and colonial wars therefore worked in tandem, producing a universal social war against autonomous being-in-the-world. Enclosure, once limited to feudal manors in England, rapidly became a geopolitical, civilizatory force. Whether cottager or aborigine, a universal alienation gripped the planet's newly anointed surplus populations.

Contemporary Enclosure

Enclosure remains an ongoing biopolitical project to capture, discipline, and regulate life. As well as being a space of control—a disciplinary force operating upon the landscape—it is also a psychological project. As Hodkinson writes, "In other words, primitive accumulation and enclosure are not just about closing off soil and land in a narrow sense but shutting down access to any space of sociality that threatens our ideological or material dependence on capitalist social relations"[91] While enclosure is undoubtedly a foundation for economic accumulation, it must necessarily secure the very contradictions it generates—namely, system-wide alienation—through physical control, intrusive surveillance, and widespread psychological conformity. This is the grim legacy we have inherited from the fencing of the commons so many centuries ago. It's the same legacy that compels drone warfare into new, ever more invasive and intimate spaces. As Linebaugh observes, "To Marx's letters of blood and fire we must now add the bomb and the drone as means of expropriation."[92]

The continued existence of the human species across the globe has seen a slow decline in the spaces of open living. Enclosure in the twenty-first century has become, increasingly, an atmospheric form of control. Crucially, this means many of the entrenched problems humanity faces are spatial or ecological questions: not simply a matter of *who* we are but *where* we are. Like a twisting Möbius strip, social and psychological relations are connected to the environments in which we live. "Henceforth," writes Félix Guattari, "it is the ways of living on this planet that are in question, in the context of the acceleration of techno-scientific mutations and of considerable demographic growth."[93] Moreover, the landscapes that surround us are constantly being remade into artificial colonies that both express and protect the planetary order of things. Shopping malls and prisons are two sides of the same coin—synthetic enclosures that capture the excess circulations of life.

Our mundane survival on this planet is now utterly policed by the rhythms of enclosure. Walking the streets of our modern cities, it's impossible to miss these apparatuses of capture that fortress and divide urban populations. As Edward Soja writes, "Not only are residences becoming increasingly gated, guarded and wrapped in advanced security, surveillance, and alarm systems, so too are many other activities, land uses, and everyday objects in the urban environment, from shopping malls and libraries to razor-wire protected refuse bins and spiked park benches designed to stave off incursions of the homeless and hungry."[94] A paranoid nervous system fuses us together inside a hyperactive technological civilization obsessed with mitigating risk. We live under the most complex and pervasive apparatuses of surveillance, enclosure, and killing in human history.

Instead of a plurality of place-specific, community-engineered worlds, the global age has installed a vast Crystal Palace. Sloterdijk takes this metaphor from the real Crystal Palace, a London spectacle opened in 1851 by Queen Victoria that hosted the Great Exhibition. This enormous, one-million-square-foot greenhouse displayed the wonders of the Industrial Revolution and British Empire to hundreds of thousands of paying visitors. In this glass construction, Sloterdijk sees a symbol for what he calls the "world interior of capital": an enclosed space that is protected, comfortable, and posthistorical. If history is a form of unilateral action, a one-way perpetration of some sort—as with European colonization—then the Crystal Palace denies its dwellers that criminal freedom. Instead, there is dome-estic bliss, a life stripped of any challenge or political strife: a shiny, balmy enclosure so spacious one might never have to exit the inside. As he explains:

> The capitalist world palace—the ultra-late Marxists Negri and Hardt recently re-measured it under the name 'empire,' albeit leaving its outer boundary deliberately unmarked, presumably to invoke more effectively the chimera of an organic alliance between the outer and inner opposition—is not a coherent architectural structure; it does not resemble a residential building, but rather a comfort installation with the character of a hothouse, or a rhizome of pretentious enclaves and cushioned capsules that form a single artificial continent.[95]

Indeed, human existence has been rendered increasingly artificial by capitalism, whose endgame is to create a kind of space station here on earth—that is to say, a completely augmented, prosthetic environment. According to the Midnight Notes Collective:

> Capital has long dreamed of sending us to work in space, where nothing would be left to us except our work-machine and repressive work relations. But the fact is that the earth is becoming a space station and millions are already living in space-colony conditions: no oxygen to breathe, limited social/physical contact, a desexualized life, difficulty of communication, lack of sun and green . . . even the voices of migrating birds are missing.[96]

To be enclosed is to be imprisoned within sociopsychic spheres: trapped on the inside, monitored by watchful eyes, and threatened by the global war machine. Our thoughts, emotions, and mobilities are enveloped by the infrastructures of technological civilization. As Hodkinson argues, "Here we can think of enclosure as imprisonment, as the enclosure of our minds and bodies within the capitalist-imperialist-authoritarian machine."[97] Contemporary enclosure thus unites neoliberalism and militarism, fusing the global market with the global war on terror, the electronic battlefield with electronic globalization, placing it all "inside" a technologically saturated civilization. Insides and outsides continue to fracture and fragment the dwindling public commons.

The being of our historical epoch, in the great age of electronic globalization, where capital, people, and data flow together like electrons on a circuit board, is nonetheless one of the most profoundly enclosed regimes in history. As Alain Badiou argues, "The price of a supposedly unified world of capital is the brutal division of human existence into regions separated by police dogs, bureaucratic controls, naval patrols, barbed wire and expulsions."[98] Boundaries, borders, insides, and outsides are cartographic and existential barriers that still segregate human beings in the electronic zoo—united in our separation. As Slavoj Žižek argues, "What increasingly emerges as the central human right in late-capitalist society is *the right not to be harassed,* which is a right to remain at a safe distance from others."[99]

Sometimes, enclosures are easy to identify: walls and wire, guards and gates. Other times, they are more difficult: drones in the sky, listening posts in the desert, satellites in space, and data storage facilities in the middle of nowhere. "If enclosure produces specific spatialities of inclusion and exclusion," writes Jeffrey et al., "these spatialities are also constituted by an *apparatus of biopolitical capture.*"[100] These apparatuses of biopolitical capture have changed with the increased technical capacity of the state to produce a vast archipelago of confinement. As Blomley writes, "The paranoia of the seventeenth-century yeoman, behind the double hedge, compares to the anxieties of the modern property owner, secure (yet always and ever insecure) behind razor wire, alarms and armed response signs."[101] Put differently, where hedges once mediated the spaces of enclosure, advanced machines now do the work of securitization. These infrastructures police the lifeworld, dividing and separating the human species in a process that integrates software and hardware, flesh and bone, economics and geopolitics.

Physical barriers are still used, of course. The 1867 invention of barbed wire was "a means of enclosing cheaper, speedier, and nastier than any other."[102] But increasingly, the enclosures that saturate cities and battlefields alike are policed by automated machines: robots, drones, computers, software algorithms, and CCTV cameras. These technologies hack into the more intangible spaces of the lifeworld: tracking, digitizing, and preempting the daily rhythms of human coexistence. And these biopolitical apparatuses are stretched far and wide, forming an artificial skin that covers much of the planetary body, smothering it with artificial colonies of all shapes and sizes. A fortress of insiders is protected from the threat of marauding outsiders. Sovereignty is this act of dividing, or separating, life in order to protect it.

This is an "immunizing process," according to Roberto Esposito, one that is always spatial and always productive of new enclosures.

Although the symbol of enclosure may have switched from hawthorn hedge to Predator drone, a civilizing mission still works upon the landscape. But this should not imply the creation of a single container. The enclosures we live in today are a shifting topology of insides and outsides that can change shape without ever losing their carceral function. Technological civilization does not create a single dome on earth but—to paraphrase Sloterdijk—installs a rhizome of suffocating enclaves and carceral capsules that form an artificial continent. The military, the police, and the various intelligence services all work to bring the dangerous outsides of the planet into a network of watched, biopolitical insides. The result is a form of state power that is centralizing without having a center: a vast civilizatory chamber that splinters into a honeycomb of insides. The Crystal Palace transmutes into a glass labyrinth, an opaque maze.

The spatialities and materialities of U.S. military enclosure exploded during the war on terror, infesting the intimate moments of everyday life, as well as staging some of the most theatrical acts of imperialism. These infrastructures are vast, tapping undersea cables, orbiting the earth, and storing terabytes of personal data. The war on terror is not reducible to the historical act of English enclosure, of course, but it takes place across the existential landscape it continues to structure. The creation of an exiled class of people, a growing surplus population, continues to generate systematic animosity. More often than not, these contradictions simply generate more invasive apparatuses of biopolitical capture. Few places are left to hide in the new enclosures. You may run, of course, but the shadow cast by the Predator Empire is never far behind.

The Production of Imperial Space

Every empire fights for space, creating a lived, policed, and pacified network of enclosures on the surface of the planet. To enclose the globe, a system of abstract geography must first be installed, both imaginatively and materially. For millennia the globe has been mapped and remapped according to the designs of cartographers: calculated, controlled, and conquered. The Predator Empire itself emerges from the historical production of imperial space. This is an abstract, calculative, and homogenizing spatiality that brings distinct lifeworlds inside its geographical matrix. The unique time–space milieus of different cultures, tribes, nations, and families are enclosed within a universalizing—and ultimately empty—kind of space.

Geographers have long been complicit with imperial projects, beginning with the institutionalization of geography as a form of state power-knowledge in the British Empire during the 1870s.[103] The U.S. brand of geographical intelligence began during World War II and the early Cold War, when "the entire Earth became a generalized space of American military strategy," according to Trevor Barnes and Matthew Farish.[104] The point here is not simply that the planet was observed by geographers but that it was remade into a set of imperial geographies. What makes past and present imperialism possible is such abstract enclosing of the planet. Geographical metaphors and imaginations have played a pivotal role in conceptualizing the war on terror, both for state elites and domestic audiences. Under the Bush administration, for example, the terrorist threat was located in a transnational "arc of instability" or an "axis of evil." This kind of geographical framing is an inherently disciplinary force in the world. "The frame," writes Judith Butler, "does not simply exhibit reality, but actively participates in a strategy of containment, selectively producing and enforcing what will count as reality."[105] Under the Obama administration's program of targeted killings, the idea of hostile containers has been overrun by a more seamless, amorphous battlespace, under what Derek Gregory calls an everywhere war.[106] State power is, then, unavoidably cartographic: its efficacy depends on a performative framing of the world.

The idea that space is produced is central to Henri Lefebvre. The challenge he lays before us is to see space and society as relationally constituted. For Lefebvre human beings operate in the grit and dirt of the planet to erect a social space to inhabit, and in turn, this social space operates on its inhabitants, colonizing their activities and thoughts. As Lefebvre argues, "The living organism has neither meaning nor existence when considered in isolation from its extensions, from the space that it reaches and produces, i.e. its milieu, every such organism is reflected and refracted in the changes that it wreaks in its milieu or environment, in its space."[107] Lefebvre's concern was, above all, how capitalism encloses and reformats space. Indeed, the very survival of capitalism depends on the continual production of space. And the space that capital produces is defined by its universality: it is beholden to instrumentality, profitability, and exchangeability. In order to link disparate parts of the globe together for profit, the world must speak the same language of an abstract space.

During the hundreds of years that defined the zealous expansion of European empires, the lifeworlds of millions of people were remapped according to "the indifferent view of things in terms of homogenous space, homogenous time and homogenous value."[108] From 1492 onward, all habitats on the earth gradually became points of equal value. "The primary fact of the Modern Age," concludes Sloterdijk, "was not that the earth goes around the sun, but that money goes around the earth."[109] By trading commodities around the planet, "every empirical place on the earth's surface becomes a potential address of capital, which regards all points in space in terms of their accessibility for technical and economic measures."[110] Floating on top of a new imperial cartography, capital swam across the newly conquered waterworld with increasing density and alacrity.

In turn, with increased ocean traffic European seafarers came to rely on a range of cartographic tools to secure the planet for the safe circulation of capital. The map was essential. "The master," writes Michael Serres, "is always a geometer, a topologist, and someone who knows space first of all."[111] The British maritime community pioneered the modern management of the seas in the nineteenth century. As Michael Reidy argues, "The machinery of empire required as its lubricant a science of the sea."[112] As he continues, "These experts transformed the vast emptiness of the oceans into an ordered and bounded grid."[113] Not only did mapping bring a sense of superiority to European cartographers, but the belief that sailors were floating on a vast outside created a space of moral abandon. Removed from the inside of their (religious) semiospheres, European seafarers soon became pirates and slave traders, criminals of the waterworld. "The other, viewed as a body in the external space, is no cohabitant of a shared lifeworldly sphere."[114] This vast externality, the hellish outside, was an *extralegal* nowhere.

By the twentieth century no point on the earth's surface "could escape the fate of becoming a location."[115] For Martin Heidegger "the fundamental event of the modern age is the conquest of the world as picture."[116] Maps, charts, and diagrams were all part of a universalizing scientific plan through which the planet could be measured. The idea of territory as a kind of calculated enclosure emerges from this matrix of imperial space. Territory is the creation of a bounded inside from a vast imperial space and is therefore a crucial political technology. As Dalby observes, "The assumption of political spaces as containers, as stable areas that are violated in some sense by the boundary-transgressing phenomena, maintains the assumption of absolute space precisely demarcated as the basis from which political discourse takes place."[117] In order to be able to contain a space, then, it must first be measurable. Therefore, "a properly critical political theory of territory," writes Stuart Elden, "needs to investigate the quantification of space and the role of calculative mechanisms in the commanding of territory, and the establishment of borders."[118]

It is no coincidence the rise of mapping paralleled the development of arithmetic and statistics in Europe. In order to carve up the planet, a spatial language that could be understood by various European nations was required. As Dalby stresses, "Above all, thinking seriously about empire requires a recognition that they are always about more than simple territorial control."[119] The development of such an imperial geometry was partly inspired by René Descartes, who saw beings as inherently calculable and measurable. For Elden, then, territory is not merely a political way of conceiving land but the result of a geometric understanding.[120] As he argues, "Late capitalism extends the mathematical, calculative understanding of territory to the entire globe. . . . Since the seventeenth century, the predominant ontological understanding of the world has been its calculability."[121] The very existence of a border, a state, or a nation is based on this a priori system of calculated land. Far from the silent background of state sovereignty, then, territory is an applied planetary discipline and a constitutive dimension of struggle. Empires are never simply territorial arrangements—their power swells from the geometric force they engineer into the world. Places become lonely points on a map, united by the impulses of money and war.

Separation and fragmentation provide the building blocks of modern political spaces and subjectivity. This puts the struggle for space as the foremost struggle waged by empire. To conceive of imperial space is to consider the homogenization of the planet, the rendering of the earth into an abstract matrix of economic points, commensurability, and generalized exchange. According to Sloterdijk, such a form of globalization "reduces all local particularities to the common denominator: money and geometry. It breaks open the independently growing endospheres and takes them to the mesh grid."[122] Using this constellation of points, an empire stitches together a world for itself, manufacturing interlocking territories of control.

This geographical game of connect the dots is still written in the landscape today. Modern Africa is divided by the longitude and latitude of past crimes: geometric scars that are testament to the work of imperial space. The globalization of calculable points has created a vast imperial space on the surface of the earth, a process of planetary-wide alienation that has only accelerated with the growth of technological civilization. Imperial space remains the generic coordinate system for world politics. Its tenacity and strength lies in its mathematical rationality. Indeed, imperial space is the necessary means for globalization, the deworlding and reworlding of the planet into new enclosures: insides that are, paradoxically, always outside.

The idea of abstract, calculative space remains a vital foundation for modern warfare, which prides itself on precise targeting. Extending Heidegger's insights, Ray Chow argues that today we live in the "age of the world target."[123] This is an age in which aerial power has enframed the world as a grid of targets. As such, the drone crystallizes a widespread targeting zeitgeist. The technologies of the Predator Empire should be understood as active participants in the ongoing production of imperial space: slicing and dicing chunks of reality to render the world in their own image. Where once maps calculated and divided the world, now drones serve as watch keepers of existence, policing reality with a robot's-eye view. Drones are part of a wider apparatus that is rewiring the existential circuitry of how people, state, and space are assembled under the Predator Empire.

Ghosts in the Machine

The relationship between the nonhuman and the human—especially, the role of machines in shaping human existence—not only lies at the heart of a more-than-human geopolitics but also represents a vital relay for understanding the existential impact of the Predator Empire. In particular, the enclosure of the planet is increasingly performed by machines that generate carceral atmospheres. If machines can indeed shape technological civilization, it's because they mediate the most intimate spaces of daily life.

Gilbert Simondon is an early pioneer of the philosophy of technology who writes that "human reality resides in machines as human actions fixed and crystalized in functioning structures."[124] Human needs have become increasingly shaped by the technical object, "which thereby acquires the power to shape a civilization."[125] As a machine becomes more self-sufficient, more concrete, it draws in the environment around it like a tree planting its roots in the ground, producing an enveloping technogeography. The machines of the Industrial Revolution, for example, permanently transformed the physical landscape of Britain. Canals and railways served as arteries, pumping raw materials into the frantic heartbeat of the factory. The Industrial Revolution didn't simply add machines to a preexisting society. Instead, the revolution changed who we were as a biological species. With the Industrial Revolution and the enclosure of the commons, humans were fundamentally deworlded and atomized such that the individual became the sole unit of social, economic, and political reality.

Marx understood machines as means of production that organized social and economic relations. The problem for Marx was not machines per se but how they were employed by capitalists. Machines, for him, were vampiric and robbed humans of their innate creativity as the worker became disciplined by the mindless rhythms of factory life. The more automated the machine became, the more the worker was reduced to the role of a watchman or overseer. As Amy Wendling writes, "In technological alienation, human beings are not only dominated by the commodities they produce; the very tools with which human beings labor dominate them."[126] Marx makes a clear distinction, however, between human and machine. He does not, according to Wendling, "extend this philosophical anthropology to its historical consequences: that the embodiment of different forms of tools produce different types of human being just as the use of different instruments produces different society."[127]

Capitalism was the emerging mode of production during Marx's life, one that depended upon the anterior enclosure of the commons. While Marx typically understood the mode of production through an economic lens, it can be understood in a deeper sense. *Capitalism created a machine-mediated reality.* Wolfgang Schivelbusch, for example, argues the advent of the railway refashioned space and time in industrializing England.[128] Owing to the necessity of synchronizing railway journeys in the 1840s, individual English railways adopted Greenwich Mean Time as a national standard. With the advent of the railway, the human species gradually came to accommodate the machine's temporal world. As Wendling writes, "With the advent of the railway as a technical and theatrical social agent, the human being must accommodate the machine's temporal world. Machines highlight, insist upon, and distribute a rhythm and temporality to the modern world."[129] The train, together with the millions of components connected to it, opened up a new technogeography, revealing a historically new mode of being.

Machines operate on the world to produce and store a distinct technogeography. This technogeography, which can be thought of as a type of artificial sphere, creates a powerful momentum that human beings must subsequently accommodate. Machines, in other words, assemble human and nonhuman forces to generate new spheres of human coexistence. Reality is, then, not a single, homogenous background but a multitude of spheres constantly shifting shape. As William Connolly explains, "The planet, on the reading advanced here, is not holistic; it is not a mere environment or setting either. It is the site of interacting, partially self-organising force-fields with variable capacities to morph."[130]

In other words, machines break down the planet into unique spheres of human coexistence. Living inside these augmented terrariums, human beings are constantly reconditioned by the push and pull of their technical creations. This process is what Simondon terms "transduction," whereby humans and nonhumans are brought together in novel constellations—this is a process of translation, or mediation, in which materials are crystallized into new milieus. As Elizabeth Grosz explains, "Transduction generates the creative leap from the past and present of the pre-individual to the unknown future, as well as fields, regions, regimes which surround and enable the being in and as its milieu. It thus generates its own kind of temporalizations and spatializations (perhaps even colonizations)."[131]

While Simondon discusses how technology produces technogeographies, the idea of transductive colonizations accentuates how machines can be invasive forces—ecologically, socially, and psychologically. The machine produces a colony, or a sphere of political sense. Within this locally stable sphere of operations, humans adapt to their changing ecology. "Collective relations are largely mediated by technical objects which elaborate and contribute to psychical cohesion," argues Grosz.[132] In short, spheric transduction *makes sense*: it makes matter meaningful—whether physical, biological, or mental. The human subject thus emerges on the inside of these spheric transductions (which means the technologies within these spheres are crucial to human existence). The human being *becomes* human through a series of technical operations, or what Simondon calls "individuations." As Muriel Combes explains, "At stake for Simondon is showing that individuation is primarily an operation."[133] This is another way of saying, once again, that reality is transductive, that the ground of existence—even psychic existence—is not an unflinching essence but an operation that crystallizes forces from across a more-than-human terrain.

As Michel Foucault elaborates, "It is therefore, I think, a mistake to think of the individual as a sort of elementary nucleus, a primitive atom or some multiple, inert matter to which power is applied, or which is struck by a power that subordinates or destroys individuals."[134] Individuals are always an effect of power, a "power-effect." The human psyche shuttles between the living and the nonliving, the individual and the collective. Bernard Stiegler names the inorganic background of human existence "technics." Perhaps, then, in the age of the Anthropocene, the earth is composed of not only a biosphere, hydrosphere, atmosphere, and lithosphere but also a technological world sphere that intersects with the planet—a kind of technosphere. "Life may well be a geological force," argues Nigel Clark, "but humans too function as earth processes, contributing to changes that alter the overall condition of biological life on the planet."[135] This vast technosphere does not simply add to the human from the outside in but is constitutive of our actions, thoughts, and feelings. For Stiegler what makes the human unique is that it exists outside itself, among technics. At birth we inherit not just a genetic memory pool in the form of an internal DNA code but also a technical memory in the form of a shared, external technics—an epigenetic apparatus.

Humans and their vast inventory of civilizatory tools invent each other—so that the *who* of the human species is also a *what*. In Stiegler's words, "If the individual is organic organized matter, then its relation to its environment (to matter in general, organic or inorganic), when it is a question of a *who,* is mediated by the organized but inorganic matter of the *organon,* the tool with its instructive role (its role *qua* instrument), the *what*. It is in this sense that the *what* invents the *who* just as much as it is invented by it."[136] Technics, the inorganic background to human life, continually transforms the social, biological, and political domains of human existence. It is, in short, the reference system of the world, writing our grammars of habit and coexistence. Our exit from the animal kingdom into technological civilization is not a giant leap for mankind: it's a thud into the arms of a planetary technosphere. Since we always exist outside ourselves in synthetic spheres, our destiny is bound to these prosthetic infrastructures that mediate the human species.

The electronic technosphere that now covers the planet is best understood as the erection of a civilizatory prosthetics that conditions how we communicate, how we live, and how we die with each other. Machines serve as watchdogs of existence wherever they operate—patterning the globe with standardized routines and mass-produced dispositions. Machines are regulatory, disciplinary even, policing the spheres of human coexistence. In this sense, to be is not simply to be-with-machines but to be-dominated-by-machines. As Gilles Deleuze wrote, "The point is that human forces aren't on their own enough to establish a dominant form in which man can install himself. Human forces . . . have to combine with other forces: an overall form arises from this combination, but everything depends on the nature of the other forces with which the human forces become linked."[137] We inhabit machines like the plants and animals we wrap around our skins. Each successive human generation inherits the weight of thousands of machines that have successfully reprogrammed human psyches. And if, to recall Stiegler, the what invents the who, then we must never stray far from the fundamental question of what kind of people technological civilization is inventing. Amid the drama stirred by drones and the excitement fueled by robotics, this existential inquiry—as old as thinking itself—must persist.

Electronic, automated, technological civilization provides the vectors for our machinic subjectivity. It is the machine that displaces or, rather, disburdens the modern human subject from the necessity of making a decision. Everywhere, the psychic confusion of confronting the world is tranquilized with the aid of technics—existence is relieved of its terrible freedom by the artificiality of modern spheres, which are nothing other than life-support systems for electronic technics run amok. These developments are creating what Guattari calls a pervasive "remote-controlling of human individuals and groups."[138] It is here the remote-controlling of humans should be understood in its fullest: as the hacking, remapping, and manipulation of psychosocial spheres by technological civilization, which has infiltrated the deepest recesses of our being. If there is a capitalist subjectivity, then it is generative of tectonic-sized hordes of consumers, "manufactured to protect existence against any event intrusive enough to disturb and disrupt opinion."[139] A society of drones.

In short, reality is the product of machinic transduction. Instead of defining reality as a passive background, what if we imagine it in its restless multiplicity: built and broken by multiple machines and housed within multiple spheres? A sphere is generated by the force fields, or the affective auras, that emanate from human and nonhuman forces. Reality, then, is always already constructed and is always splintered. It is in this sense technology must be thought of as existential, since its operations create new spaces of dwelling. As beings-in-spheres, we feel these operations deep in our bones. Whenever the materiality of the world is redistributed into new modes of existence, our thoughts and feelings are reformatted. Living inside our prosthetic spheres, the human species is transformed by its nonhuman neighbors. For these reasons the Leviathans of the Predator Empire augur a series of existential shockwaves. Dwelling inside a booming Droneworld will fundamentally condition who we are as a people, as a society, and will alter the exercise of state power and international relations. These robotic creations insist upon us, bending new—and at times terrifying—space-times around their orbits.

The Open Prison

The question of who we are—and where we are—has an urgency amplified by the looming planetary drone wars. But rather than being understood as a strange aberration, drone warfare must be seen as a perfectly rational form of slaughter for a technological civilization devouring itself. Human coexistence is overrun with apparatuses that survey, control, and discipline the population. Agamben, working from Foucault's concept of the *dispositif,* defines an apparatus as "literally anything that has in some way the capacity to capture, orient, determine, intercept, model, control, or secure the gestures, behaviors, opinions, or discourses of living beings."[140] From this broad definition, he argues there is a massive "partitioning" of beings into two classes—living beings and the apparatuses in which they become captured. "It is clear that ever since Homo sapiens first appeared, there have been apparatuses; but we could say that today there is not even a single instant in which the life of individuals is not modeled, contaminated, or controlled by some apparatus."[141]

Outside the apparatus lies the Open. This is the infinite plane of possibility from which we build our finite worlds. A range of apparatuses infects the Open with "instruments, objects, gadgets, odds and ends, and various technologies."[142] In other words, the Open is always captured, and as Agamben notes, this is the power of the apparatus: to produce "a separate sphere."[143] Disciplinary apparatuses endlessly manufacture spheres for us to live within: channeling life away from the possible and into the merely predictable. As Agamben concludes, "Apparatus, then, is first of all a machine that produces subjectifications, and only as such is it also a machine of governance."[144] The apparatus is a form of power with no human face, a systematic or objective form of violence. As Slavoj Žižek explains, "Systematic violence is thus something like the notorious 'dark matter' in physics, the counterpart to an all-too-visible subjective violence."[145]

In turn, the Open is coming to resemble an open prison. Writing in 1954, Ellul argues a civilization overrun with apparatuses is leading us toward a worldwide concentration camp. By this, Ellul argues our lives increasingly take place inside carceral spheres. "The Nazi's use of concentration camps has warped our perspectives,"[146] argues Ellul, since the general logics of the concentration camp—endless surveillance, administration, and ordering techniques—have become diffuse apparatuses of society. Only instead of watchtowers, barbed wire, and guard dogs, there exists a vast archipelago of surveillance enclosures. As Ellul insists, the police are enclosing society with a "technical apparatus" that could be applied "everywhere" and to "everyone."[147] Crucially, our daily lives have become recordable by objective rather than subjective means. Nonhuman mediators rather than individuals have increasingly come to police society. It is these techniques of objective control, writes Ellul, that increasingly create "a milieu, an atmosphere, an environment, and even a model of behavior in social relations."[148] A widespread climate of conformity is produced by ubiquitous surveillance.

Objective, or automated, techniques neglect all individual differences and capture an immense volume of living things in multiple electronic dragnets. It is in this respect we must understand the Predator Empire. Classically, authority hinged on the power to terrorize, often with dramatic acts of public violence. But this no longer defines the secured atmospheres of the open prison. State power is instead a technical form of power that absorbs the minutiae of life: a series of interlocking apparatuses that erect a carceral technosphere with blurred edges. An extraordinary depoliticized expert infrastructure rules over us—creating "a blind faith in the technocrats of the State apparatus."[149] If politics was once the pursuit of progress for humanity, one that posed questions to the Open, it now asks questions only of the enclosed interior—namely, how do we construct an efficient, comfortable, and secure prison? More than ever, the human subject is trapped by its very fragmentation: tossed inside the gravitational pull of multiple spheres, choking on the stale air of enclosure. And there's no easy way of rewinding the clock of technological civilization. "Enclosed within his artificial creation," Ellul laments, "man finds that there is 'no exit'; that he cannot pierce the shell of technology to find again the ancient milieu to which he was adapted for hundreds of thousands of years."[150]

The Leviathan, on the face of it, protects the human species from the brutal war of all against all. Dangerous human instincts are internalized in the name of a greater good and are carefully managed by a system of advanced technics. But passions do not simply disappear. As Sigmund Freud explains, "Civilization expects to prevent the worst atrocities of brutal violence by taking upon itself the right to employ violence against criminals."[151] Crucially, this means violence is not erased by the Hobbesian sovereign—rather, it is centralized within the flesh and bones of the Leviathan. This fossilizes a deep and pervasive social war in everyday life. How, then, can a civilization that internalizes such a social war survive? In the first instance, the mind becomes the target of mental enclosure. "Civilization therefore obtains the mastery over the dangerous love of aggression in individuals," writes Freud, "by enfeebling and disarming it and setting up an institution within their minds to keep watch over it, like a garrison in a conquered city."[152] Love, hate, and lust: a raft of emotional states bubble beneath the surface of the great civilizatory spheres.

Religion, for millennia, has checked the system's pathologies, serving as a symbolic watchman in the minds of individuals. Our gods today, however, have fallen from the heavens and become soldered in the circuit boards of technological civilization. So while technological civilization immunizes its pathologies no less than any other civilization in history, the religious apparatuses that once policed the system's illnesses have slowly lost their grip on the species. These symbolic shells began to evaporate with modernity. As

Roberto Esposito argues, "It is this tear that suddenly opens in the middle of the last millennium in the earlier immunitarian wrapping that determines the need for a different apparatus of the artificial sort that can protect a world that is constitutively exposed to risk."[153] Artificial technics must now defend, police, and discipline a growing and increasingly restless surplus population, everywhere securing the insecurities of enclosed living. The Predator Empire is thus the immunitary modality of technological civilization. Its drones ensure we remain locked inside our civilizatory enclosures, comfortably numb.

2

The Rise of the Predator Empire in the Vietnam War

Ever since the bow and arrow was invented thousands of years ago, remote killing—in some form or another—has been a defining feature of war. This chapter investigates a key period in this long history: the Vietnam War (also known as the Second Indochina War or the American War). This conflict, which officially took place between 1965 and 1973—but expanded well beyond these dates—created a series of important precedents for the Predator Empire of today. One of the most obvious lines of descent is that the Vietnam War birthed the most sophisticated program of drone surveillance in the history of flight. According to James Gibson's influential analysis, the Vietnam War was history's first technowar: a war conducted according to technical principles, statistical models, and machinic systems.[1] Of particular importance was the rise of the electronic battlefield. During the 1960s the U.S. Department of Defense began to automate and computerize the battlefield with remote sensors and supercomputers.

The Vietnam War was a technologically intensive conflict fought with sophisticated electronic prosthetics, from remote sensors that listened to enemy movements to jet-powered Firebee drones that screamed through the skies. A gigantic electromagnetic dome was slowly erected by the U.S. military over Vietnam, Laos, and Cambodia. In its interior an artificial world was installed: a security sphere managed with the certainties of cybernetics and stalked by teams of manhunters under the Phoenix Program. The tropical atmosphere bounced with radio signals, laser beams, B-52 bombers, Huey helicopters, the toxic drift of Agent Orange, and the smoke plumes of burning forests. But the dome was never entirely sealed by the U.S. military. Underground, thousands of tunnels were dug to escape from the electronic enclosure.

Taking inventory of the Vietnam War thus paves the way for a deeper understanding of the Predator Empire. And there are multiple geographies and infrastructures of violence to be considered: techno-war and U.S airpower, ecological warfare, the electronic battlefield, manhunting, and drone surveillance. These geographies converge in the security logics and practices of the modern Predator Empire—and what ties them together is the centrality of the atmosphere as both a *medium* and a *target* of state violence. In this sense, the Vietnam War is a crucial period in establishing the foundations of atmospheric warfare today. On the one hand, the U.S. military destroyed multiple lived environments across Vietnam, Laos, and Cambodia—bombing, bulldozing, burning, and poisoning the forest. On the other hand, the conflict inaugurated forms of atmospheric warfare that enclosed the biopolitical landscape by using technical apparatuses. This includes the electronic battlefield, the bureaucratic architectures of the Phoenix Program, and the aerial surveillance orbits of Firebee drones. Atmospheric warfare, accordingly, must be understood as both *ecological* and *electronic,* collapsing physical and virtual surfaces inside the artificial interiors of the war machine.

It is difficult to appreciate just how quickly military technology advanced during the Cold War. On October 4, 1957, the Soviet Union launched Sputnik 1. Propelled in the midst of the Cold War, this metallic orb triggered a billion-dollar scramble for research into electronics and robotics. The gigantic U.S. military–industrial complex was awoken. But Sputnik wasn't the only satellite the Kennedy administration had set its eyes on in the looming space age. Landing on the moon would soon become a national priority. This chalky rock has always glowed in the imaginations of earthlings, but for millennia it had been an unreachable destination. This changed on July 20, 1969, with the Apollo 11 mission. Stepping on the moon, the astronaut Buzz Aldrin described what he saw from behind his shiny black visor: "Beautiful, beautiful. Magnificent desolation."

Two years before this manned landing on the moon, on April 20, 1967, the Lunar Surveyor 3 craft became the second Surveyor robot to successfully land on the surface of the moon. Significantly, it was the first machine fitted with an extendable arm. This claw scooped at the lunar soil and broadcast the imagery back to Earth. U.S. scientists had built a prosthetic that allowed humans to touch and sense the moon's surface from a quarter of a million miles away. "Employing machines as man's extended self was nothing new as a concept, but the 1960s brought that concept to true, sophisticated reality," writes Paul Dickson, "or, at least far enough along to allow him to put his hand on the moon—a feat that for the public was lost in the excitement of the manned moon landing."[2] Back on Earth, the idea of using electronic prosthetics for sensing and manipulating the environment would explode in the Vietnam War.

On October 14, 1969, months after Neil Armstrong and Buzz Aldrin's moonwalk, General William C. Westmoreland addressed the Association of the United States Army in Washington, D.C. By this time Westmoreland was no longer in charge of U.S. military operations in Vietnam. His war of attrition in Southeast Asia had largely failed. But that was not the topic of his talk. Instead, he spoke of a "quiet revolution" in Vietnam that was rewiring and automating the battlefield, replacing "man with the machine." As he predicted, "On the battlefield of the future, enemy forces will be located, tracked, and targeted almost instantaneously through the use of data links, computer assisted intelligence evaluation, and automated fire control."[3] His prophecy was an early glimpse into the kind of drone warfare that would emerge at the end of the century, partly as a result of the technological leaps made during the Vietnam War.

Sophisticated sensors, unmanned drones, automated machine guns, laser-guided bombs, computer algorithms, mobile attack helicopters, and the cybernetic systems that held them all together were largely unprecedented in the history of U.S. warfare. So many superlatives were spilled the newspapers could barely keep pace. War managers have always sought, of course, to command and control reality, battling against the fog of war that clouds with their predictions. It was during the Vietnam War, however, that the promise of a crystal-clear battlespace—one in which the fog of war would evaporate entirely—was first promised by the technocrats. But this computational certainty was contradicted by slaughter. A mass of American soldiers was thrown against a guerilla army that knew the land better than they did. Karl Malantes—decorated Vietnam veteran and author of *Matterhorn*—captured the madness when he wrote, "People who didn't even know each other were going to kill each other over a hill none of them cared about."[4]

The Colonial Context

On April 30, 1975, the capital of South Vietnam, Saigon, was evacuated in chaotic scenes played on television screens across the world, marking the "end" of the Vietnam War, which had raged for two years after President Nixon withdrew U.S. troops. U.S. ground intervention began March 8, 1965, following the Gulf of Tonkin incident a year earlier, but special forces and military advisers had been in the country since the late 1950s. Indeed, the Vietnam War had no clearly defined beginning, let alone a distinct end. Accordingly, weaving a single narrative about the conflict is difficult. So while this chapter is organized around a series of spiraling technogeographies, it's important to begin with the colonial legacies to the conflict. It is crucial to understand the Vietnam War was never only an *American* war. It reached back decades, centuries even, long before U.S. Marines deployed onto the beaches of Da Nang.

French missionaries had been present in the country from as far back as the 1600s. In the nineteenth century France's colonial grip on Vietnam strengthened after China's influence across the region was weakened during the Opium Wars. France created what was known as French Indochina, composed of the Vietnamese regions of Cochinchina, Annam, and Tonkin, as well as Laos and Cambodia. The French imposed land reforms that converted vast swathes of the countryside into rubber plantations, particularly in Cochinchina. "Sometimes benevolent, but often brutal overlords, the French more or less attempted to administer Vietnam for profit."[5] Against this French rule, a swell of unrest churned, with both Annam and later Tonkin becoming hotbeds of a growing anticolonial struggle. Several nationalist parties were formed, with the Communist-aligned Viet Minh, created in 1941 and led by Ho Chi Minh, becoming the most tenacious voice in the struggle to shake off decades of colonial rule. The Viet Minh eventually gained de facto rule of Vietnam, but not before an intervening period of Japanese occupation during World War II (during which Vietnam received U.S. support). France then sought to retake its colonial possession in the early 1950s, which was where the roots of the modern conflict—and U.S. involvement—began.

The French military's occupation of Vietnam in the 1950s foreshadowed U.S. military strategy. By dividing the country into a lattice of roads and forts, the French military attempted to convert the tangled mess of the forest into a disciplined imperial space. Seduced by their technological superiority over this abstract plane, the French believed that "the more mechanized war machine would surely destroy the primitive one by virtue of the technological superiority of one 'machine' over another."[6] But this technological conceit was undone. The French met decisive defeat at the infamous battle of Dien Bien Phu in 1954. The subsequent Geneva Accords divided the country along its seventeenth parallel, with a Ho Chi Minh–administered territory north of a French- and U.S.-supported south. Elections in 1956 were mandated to reunify the country. But the ballot boxes never came. The United States—which had underwritten France's reoccupation—backed the anticommunist Ngo Dinh Diem regime in the South, fearing a landslide Communist victory. Violence subsequently erupted as Ho Chi Min's People's Army of Vietnam (PAVN) clashed with the South.

Instead of unifying Vietnam, the United States had thrown its support behind an administration, the Republic of Vietnam, that had little popular support (not to deny the various social, religious, and territorial complexities throughout the country). Perhaps unsurprising, Diem aligned himself with Vietnam's landlord class, rolling back Viet Minh land reforms and gaining the reputation as a modern-day emperor. Local opposition gained momentum, and in the 1960s the National Liberation Front (NLF) was established. This insurgent group gained the name Viet Cong, or VC to the Americans (and much of the English-speaking world). The organization of these guerilla fighters was not as spontaneous as thought at the time: many of the core members were professionally trained in the 1950s by the North. According to Colonel Bui Tin, "Throughout the war, Hanoi disguised the fact that the NLF was controlled from the North."[7]

With conflict erupting between North and South, the United States—first under John F. Kennedy and then under Lyndon B. Johnson—increased its support of the Diem regime in the early 1960s, sending advisers, special forces, and other war matériels. During this period the South alienated many peasants by removing them from their ancestral homes and relocating them to so-called Strategic Hamlets—what often amounted to concentration camps. "We all know that a person can only die once," said two Vietnamese survivors, "but in Vietnam at that time, it was possible to die twice, because living in the Strategic Hamlets was a living death."[8]

Why did the Democratic Kennedy administration choose to buttress French colonialism and then carry its bloody torch, leading to the death of over 58,200 U.S. service members? Why did the U.S. military end up spending around $167 billion against a peasant army?[9] The answer is Vietnam was perceived as a Cold War battlefield by the Washington war managers, one that needed to be crushed and contained. During this period U.S. Cold Warriors endlessly divided the world between good and evil, capitalism and communism, freedom and despotism—a moral clarity the war on terror would exhume decades later. As Andrew Bacevich insists, "Beginning with Franklin Roosevelt, every U.S. president had insisted that at the far side of America's resistance to totalitarianism world peace awaited."[10]

This narrative is, of course, misleading. Geographic particularities were erased by a Cold War clarity, and the war managers of the U.S. military were "blinded to many of the deep subtleties that existed during the conflict, especially among nations yearning to throw off the yoke of European colonialism."[11] As Nick Turse writes, "The United States never wanted to admit that the conflict might be a true 'people's war,' and that Vietnamese were bound to the revolution because they saw it as a fight for their families, their land, and their country."[12] In short, the Vietnam War was always more than a military conflict pitting North versus South: it was a colonial war, a civil war, and a Cold War. And it was history's first technowar.

Technowar

Gibson describes the Vietnam War as a technowar, "a high-technology, capital intensive production process."[13] Winning the war was understood as a technical rather than a strategic problem: "Military strategy becomes a one-factor question about technical forces; success or failure is measured quantitatively. Machine-system meets machine-system and the largest, fastest, most technologically advanced system will win. Any other outcome becomes *unthinkable*."[14] This is how the enemy was understood by U.S. war managers: as a system composed of elements. But technowar has a trajectory longer than the Vietnam War's and emerges from the industrial warfare that preceded it.

General Westmoreland was commander of U.S. military operations in Vietnam until 1968. He understood the conflict broadly as a war of attrition whose aim was to force the NLF into submission via large-scale search-and-destroy missions. This faith in the power of mass was reflected in the training the U.S. Army and Marine units received at the outset of the war. The soldiers were prepared largely for European battalion-sized conventional combat. But the NLF operated differently, following Mao Zedong's doctrine that "when the enemy advances, withdraw; when he defends, harass; when he is tired, attack; when he withdraws, pursue." This means the U.S. military was constantly fighting on terms with which it was not familiar and explains why the NLF so often defined the sites of engagement, littering the jungle with ambushes and booby traps. "As the battlefield was not linear, there were no front lines facing each other; no enemy fixed positions; no headquarters, artillery positions, supply dumps; none, in fact, of the targets found on a conventional battlefield," writes Gordon Rottman.[15]

The U.S. military's mechanized fighting was a result of the industrial warfare that had preceded it for decades. The first industrial conflicts were the total wars of the Napoleonic period, which mobilized hundreds of thousands of conscripts and dragged entire nations into battle. Napoleon's artillery units indiscriminately pounded humans into scraps of meat. Industrialization devalued war just as it devalued life: conflict became an impersonal, machinic calculation.

World War II was the apex of industrial warfare. Entire cities were bombed, burned, and annihilated across the planet. According to the International Committee of the Red Cross, for example, around one hundred thousand tons of mostly napalm bombs were dropped on sixty Japanese towns and cities.[16] The most deadly Allied attack came under Operation Meetinghouse in March 1945 when 1,665 tons of napalm were dumped on Tokyo, Japan's "paper city" capital. Over eighty-three thousand were killed as a cauldron of fire, wood, and flesh roared: "Those who fled to the city canals faced death from the scalding water or from the stampeding mob crowding in and crushing on top of them."[17] This strategic bombing was recommended by an analyst at the U.S. military's Statistical Control Office. He would come to embody the transition from industrial warfare to electronic warfare. Robert Strange McNamara was his name—and he would personify the cybernetic revolution about to sweep the Pentagon.

McNamara, a graduate of Harvard Business School, was the influential secretary of defense during the early years of the Vietnam War. As a military analyst during World War II, he used IBM machines and "human computers" to calculate low-altitude firebombing sorties "with devastating results for Japanese cities."[18] After the war he joined the Ford Motor Company and brought the same scientific management principles honed during his operations research at the U.S. military. President Kennedy picked him to be secretary of defense in 1961. McNamara, hailed as a management genius, brought a cadre of computer whiz kids and systems analysts to the Pentagon. His aim was to transform the military into a corporate system. McNamara "relied on numbers to convey reality and, like a machine, processed whatever information he was given with exceptional speed. . . . There was to be no 'fog of war' for his Pentagon."[19]

In turn, this produced a managerial approach to waging war. In 1962 McNamara began the Planning, Programming, and Budgeting System (PPBS), integrating systems analysis across the military and government and seating a technocratic elite on the thrones of power.[20] This technocratic worldview was in no small part welded to the rise of computers, which soon came to reverse engineer reality as information became the basic building block of existence for U.S. war managers. This produced a self-referential space of military knowledge. In late 1966, for example, McNamara ordered the CIA to design a system that would report on "pacification" efforts in the countryside. Intelligence analysts then used a scale from A (secure) to E (contested) to categorize South Vietnam's 12,600 hamlets. These scores were then processed and displayed on a consolidated IBM dot-matrix map: "Beneath the high-tech patina, an essentially subjective assessment pronounced 75 percent of South Vietnam's population pacified by late 1967—just before the disastrous Tet Offensive shattered such illusions of progress."[21]

The management of the Vietnam War was influenced also by the growing impact of cybernetics, the study of feedback loops in systems. A cybernetic system is composed of a sensor that monitors an environment, a processing unit (such as a computer), and an output that alters the environment. Robert Wiener, the founder of cybernetics, and others like him discovered cybernetic systems everywhere: in society, in technology, in biology, and of course, in war. As Antoine Bousquet concludes, "The cybernetic model of warfare erected by the system analysts was one that was frictionless, a perfectly oiled machine resting on elegant mathematical constructs."[22]

One statistic came to dominate all others during the Vietnam War: the enemy body count. Obtaining a high body count was key to promotion for many officers. Some established production quotas and box scores to keep a list of enemy deaths.[23] Lieutenant General Julian J. Ewell commanded the Ninth Infantry Division, which oversaw the densely populated Mekong Delta region. As with McNamara, Ewell saw the Vietnam War as an assembly line or factory that had to be organized as efficiently as possible. Ewell was criticized for being obsessed with the body count, establishing standards to determine satisfactory kill ratios. At best, this could incentivize the inflation of kill counts—at worst it rewarded indiscriminate killing. Arriving at the delta in 1968, Ewell set about creating a total war. The most prominent operation was called Speedy Express and ran from December 1968 to May 1969, during which time the military's kill ratio soared to 134 to 1.[24] As such, massacres like the My Lai incident must be understood as products of a wider system of violence.

Understood through a managerial framing of elements, systems, and geometries, the slaughter on the ground was abstracted into numbers. Rather than illuminating the world with the bright light of statistical clarity, cybernetic warfare blinded its practitioners. Eyes were wide shut. But this was not widely understood at the time, since the war managers lived in the enclosed world of the military dome. Reality was not fleshy, messy, and bloody: reality was the exchange of information. This military reason shielded the system from external thought and criticism—even as thousands of Americans died. Technowar thus overtook strategic thinking, and military reason sought to reconstruct the terrain of Vietnam as an abstract, universal space—a militarized enclosure that was knowable, mappable, and bombable.

Airpower: Commanding the Skies

Technowar relied heavily on airpower, from Huey helicopters to behemoth bombers. Airplanes created a vertical form of surveillance and destruction, installing a permanently unequal power relation between the predator in the sky and the prey on the ground. The U.S. Air Force waged an unprecedented air campaign in Southeast Asia, dropping more ordnance than was used by all sides during World War II. Between 1965 and 1972, the U.S. Air Force and the Army of the Republic of Vietnam Air Force (AVRN) flew 3.4 million combat sorties across Southeast Asia, with a plurality over the South.[25] The war managers typically understood bombing through the mediation of an abstract technogeography. As Derek Gregory argues, "Throughout the targeting process the language of patterns, areas, circles, holes and boxes erased people from the field of view; bombing became a deadly form of applied geometry."[26]

Omnipotent and omniscient as the war machine must have

appeared to its proponents, U.S. bombing was nearly always counterproductive, as scarred civilian populations often increased their support for the NLF.[27] In total, nine thousand out of fifteen thousand rural Vietnamese villages were destroyed, 1.5 million Vietnamese were killed, and 1.1 million children were orphaned.[28] Millions more people were forced to abandon their homes after the war, and most of the country's infrastructure was destroyed.

Operation Rolling Thunder was a bombing campaign designed to interdict supplies and personnel moving across the seventeenth parallel (also known as the "demilitarized zone," or DMZ) that roughly marked the separation between North and South Vietnam. Striking targets in the North was designed to psychologically shock the Hanoi war managers into early peace negotiations. The Joint Chiefs of Staff initially produced a list of ninety-four targets the navy and air force would strike over an eight-week period, commencing on March 2, 1965.[29] President Lyndon B. Johnson imposed a strict and convoluted system for target selection.[30] For technowar enthusiasts, bombing was a form of communication that conveyed a message to the enemy. But airpower was contradicted by its own conceits, as the North Vietnamese simply failed to play by the rules set in McNamara's game theory. Just before the 1968 presidential election, Johnson suspended aerial operations above North Vietnam.

In South Vietnam—without the North's aerial defenses—U.S. bombing missions had fewer restrictions, and in free-fire zones there were even fewer. Alienated, indifferent, and totalized war was not official policy, of course. For example, Directive 525-3 called for a clear demarcation between civilians and insurgents within "specified strike zones" (which were earlier termed *free-fire zones*). But in reality it was impossible to divide Vietnam into such black-and-white spaces. With 40 percent of the South Vietnamese countryside considered under Viet Cong influence, "Westmoreland's policy made that entire territory theoretically open to unrestrained attack."[31] Operation Arc Light missions, under the control of Strategic Air Command, were bombing runs delivered by high-altitude B-52 bombers. Yet these airplanes could take twelve hours to arrive after a target had been called in. Furthermore, given that ordnance was dropped from about twenty thousand feet, the B-52s' target box could be up to one mile by two miles, with many bombs often missing their targets altogether.

Beyond North and South Vietnam, the countries of Laos and

Cambodia were battered by U.S. bombs. Both governments had officially declared their neutrality to the conflict that raged alongside their eastern flanks, but both countries contained corridors for the movement of antigovernment troops. U.S. bombing of Laos and Cambodia remained covert for much of the war. The target of bombing sorties was the Ho Chi Minh Trail, which cut through both countries. Although never a fixed system of roadways and pathways, the Ho Chi Minh Trail was the main logistical network that fed the NLF in the South with resources, men, and other materiels. Indeed, major routes of the Ho Chi Minh Trail existed as free-fire zones.[32] After Johnson halted Operation Thunder, surplus aircraft were rerouted to Laos under Operation Commando Hunt air strikes, which targeted the trail and ran until 1972. Yet like a river, the Ho Chi Minh network simply shifted its tributaries. Despite the secretive nature of the Laos campaign, five hundred thousand U.S. sorties were deployed to and two million tons of bombs were dropped on the country.[33]

The campaign in Cambodia was much the same. Operation Menu commenced in March 1969, and its replacement, Operation Freedom Deal, delivered hundreds of thousands of tons of bombs to interdict PAVN forces and resources moving south. And again, by the time President Nixon pledged to halt the bombings, there was no real success to mention—no success, that is, apart from the number of people killed and the blowback the bombing set in motion. As Chalmers Johnson surmises, "President Nixon and his security adviser Henry Kissinger ordered more bombs dropped on rural Cambodia than had been dropped on Japan during all of World War II, killing at least three-quarters of a million Cambodian peasants and helping legitimize the murderous Khmer Rouge movement under Pol Pot."[34]

The PAVN launched a full-scale attack against the ARVN in the spring of 1972, which was known as the Easter Offensive. The Nixon administration responded with the first wave of B-52 air attacks on North Vietnam since Rolling Thunder had ended. The program was called Linebacker, and its follow-up, Linebacker II, which came in December, was the heaviest bombing mission since World War II. Unlike under President Johnson, far fewer restrictions were placed on these bombing missions. Although politics certainly played its part in the Linebacker missions—Nixon wanted to be seen as communicating a bullish message—the surgical precision of a new class of laser- and television-guided smart bombs was key. As the *New York Times* reported in 1974, smart bombs "fulfill the age-old military dream for a weapon so precise that one shot will destroy a target."[35] In Vietnam one airplane would typically fly as a laser designator while another would release the guided bomb.

Laser developments, instrumental to modern drone warfare, first began in the 1950s when the idea of focused electromagnetic radiation took hold. Much of the secret laser testing and experimentation took place at the U.S. Army Missile Command at Redstone Arsenal.[36] Paveway (Pave, or PAVE) was the air force's umbrella term for laser-guided munitions and precision avionic systems. In concert with Redstone, Texas Instruments of Dallas engineered laser-guided bombs in the mid-1960s. The Pave Knife was an early laser designator designed for the F-4, and by 1971, $6.3 billion was being spent on advanced laser weaponry.[37] Many predicted the laser would revolutionize warfare.[38] To an extent it did. Far exceeding the capabilities of the Norden bombsight, which in World War II increased the accuracy of dropped munitions, lasers plummeted a bomb's circular area of probability (the measure of its accuracy) from hundreds to tens of feet.

The television-guided, or TV-guided, bomb contained a sensor in its nose that broadcast visual data to a pilot's cockpit display. The pilot then manipulated the camera until it locked on to the desired target. At this point no further guidance was required—"fire and forget." The expensive TV-guided bombs were hampered by their requirement for a high degree of black-and-white contrast between the target and its environment. This was tricky in the thick canopies of Southeast Asia. But these bombs were, after all, just the beginning. As one newspaper quipped, "In fact, the 'smart bombs' now in use are about sixth graders compared with the Ph.D. bombs either on the drawing board or already being tested."[39]

One of the most widely reported uses of smart bombs came after the destruction of North Vietnamese bridges, which had long evaded "dumb" bombs. This was to be the dawn of "a new philosophy" enabled by a growing "electronic environment."[40] Despite all of the pats on the back, however, strategic bombing was never a success. If Rolling Thunder mobilized a mass and Linebacker mobilized precision, then both operations failed to understand bombs would never extinguish the workings of a highly dispersed, networked, and rural society. But failure was simply noise to technowar.

Scorched Earth: Geographical Warfare

Underpinning technowar was a violence waged directly against the environment. As Gibson explains, "Unable to see outside Technowar, war-managers attempted to destroy 'raw nature' with its mountains and forests and places to hide. They tried to create a physical terrain equivalent to the abstract, mathematical space of 1,000 meter by 1,000 meter grid squares necessary for jets and artillery to find orientation."[41] The idea was to subdue the terrain by bulldozing it, burning it, and bombing it, reconstructing Vietnam into an abstract warscape. To that end, a range of U.S. war machines operated upon the landscape of Vietnam, Laos, and Cambodia.

Arc Light and harassment-and-interdiction bombing missions wounded the landscape with devastating effect. By the end of the war, 21 million bomb craters were estimated to have scarred the South Vietnamese landscape.[42] Such craters became flooded in the tropical Mekong Delta region nearly year-round, acting as breeding grounds for mosquitos and creating a cascade of environmental effects that hampered the countryside's irrigation systems.[43] Shrapnel from exploding bombs also punctured tree bark, facilitating the growth of wood-rotting fungi.

In 1972 Yves Lacoste put forward the idea of geographical warfare in his analysis of U.S. bombing of dikes in North Vietnam. Describing it as a new phase in the history of state violence, Lacoste argues the landscape—rather than the people—was the enemy. As he writes, "For the first time in history, the modification and destruction of the geographical milieu (in both its physical and human aspects) is being used to obliterate those very geographical conditions which are indispensable for the lives of several million people."[44] Peter Sloterdijk argues, in a similar vein, geographical warfare is a defining quality of the modern age. As he writes, "The 20th century will be remembered as the age whose essential thought consisted in targeting no longer the body, but the enemy's environment."[45] But geographical warfare extends beyond bombing.

The U.S. military sprayed approximately 72 million liters of herbicide across South Vietnam, of which 42 million liters were Agent Orange.[46] These defoliation flights first began in late 1961. The U.S. Air Force modified six C-123 aircraft for the missions. These later became part of Operation Ranch Hand, which ran until 1971. By poisoning plant and tree life, the U.S. military aimed to strip the forests bare, reducing the cover for the NLF and forcing civilians to relocate. Defoliation could therefore annihilate the entire means of subsistence for rural communities. There were other consequences, too. Wildlife was often unable to survive in the new, deathly ecosystems. Other times, domestic animals were deliberately targeted, from chickens to water buffalo.[47] Residual dioxins from Agent Orange are still responsible for a number of health problems in Vietnam's population and in U.S. veterans exposed to the spraying.

Burning the landscape was another form of geographical warfare. U.S. soldiers carrying flamethrowers and Phantom and B-52 airplanes delivered approximately four hundred thousand tons of napalm during the war, with fires ravaging around one hundred thousand acres of forested land.[48] Additionally, the destruction of forests could be realized more directly. Hundreds of square miles of Vietnamese forests were bulldozed by Rome plows. These 33,000-kilogram armored tractors tore up approximately 2 percent of the total land area of the South, leaving behind a destructive ecological legacy.[49] Overall, damaged forests were estimated to account for half of South Vietnam's forested land area.[50]

The U.S. Air Force also ran a project to precipitate torrential downpours. The aim of this incipient form of weather warfare was to flood the Ho Chi Minh Trail. Hercules and Phantom airplanes released photoflash cartridges that contained silver and lead iodides inside clouds to trigger the release of moisture. "Ironically, typhoon-induced rains interfered with cloud seeding, cooling the earth and preventing the updrafts of heated air that were essential to the project."[51] This practice of climatic modification, or weather warfare, was later banned in a 1977 UN convention. In short, an armada of weapons and machines were unleashed upon Vietnam, Laos, and Cambodia, producing a widespread climate of conflict. In a geographical war the landscape becomes both the target and the space of terror.

If Agent Orange, napalm, Rome plows, and B-52 bombers represent the tools of a large-scale form of geographical warfare, then the use of tear gas represents a more intimate and atmospheric form of attack. The U.S. military first began equipping the South Vietnamese Army with so-called riot-control gases in 1962 under its Military Assistance Program. It was first used by marines during the 1965 Operation STOMP in Qui Nhon. Newspapers at the time described tear gas as "particularly useful in reducing civilian casualties when the enemy has infiltrated into population centers or built up areas."[52] Tunnels were dangerous spaces for U.S. soldiers to access, since they were frequently ambushed. "To avoid these confrontations, it soon became standard practice to throw tear gas grenades into the entrance and . . . force the gas down into the recesses of the tunnel."[53] Tunnels, of course, provided easy escape from U.S. geographical warfare. According to one 1972 magazine, "Basic training for the NVA [North Vietnamese Army] soldier lasts nine months, four of which are devoted to the practice of digging. It is estimated that without using heavy earth-moving or earth-drilling machinery, the average NVA private can dig twenty feet down in *one* hour in average soil. This extraordinary burrowing capacity vastly increases the life expectancy of enemy soldiers."[54]

One irony of the tear gas use in Vietnam was how quickly the gas was used against anti–Vietnam War protestors in the United States. On May 15, 1969, at the University of California–Berkeley's Sproul Plaza, low-flying helicopters showered hundreds of student demonstrators with plumes of tear gas while National Guard troops fired canisters into the crowd. So while the atmosphere can support life, it can also take it away: in a geographical war the atmosphere becomes the space of surveillance and terror. In Vietnam the U.S. military's will to dominate the enemy's environment reached a level far beyond the chlorine and mustard gasses first used in World War I. Industrial-scale atmospheric warfare arrived in Southeast Asia with devastating effects. This must be understood as kind of total war that draws humans and nonhumans into its deadly volumes. Thinking through these links between state violence and the atmosphere does not, however, limit us to "nature."

The Electronic Battlefield, Part I: Igloo White

While atmospheric warfare is linked directly to biological, or organic, forms of terror, our understanding of it as *atmospheric* can be broadened so that it encompasses the electromagnetic spheres that enclose and secure the circulation of life. Nowhere was this form of electronic enclosure more important than in the development of the electronic battlefield, which according to Dickson would "spark an entire generation of new weapons, change the complexion of war in Southeast Asia as well as future wars, give new direction and unity to the massive military research and development apparatus, and alter the image of war."[55] Moreover, this new sort of Manhattan Project installed important infrastructures of the Predator Empire.

The movement of PAVN men and materials along the Ho Chi Minh Trail continued to hamper U.S. aerial interdiction efforts. Unsurprisingly, technowar turned to technoscientists for new solutions. The Jason Division was a group of the nation's brightest scientists and physicists and was part of an influential Washington, D.C., think tank, the Institute for Defense Analyses. In the summer of 1966, it was charged with reviewing U.S. military strategy in the Vietnam War. The Jasons proposed constructing a gigantic fence across the southern belt of the DMZ and into the Laotian panhandle. This ambitious scheme went by various names: the Jasons called it the Air Supported Anti-Infiltration Barrier; the Senate Armed Services Committee named it the Electronic Battlefield as well as Igloo White, after the main segment of the network that cut through Laos; and finally, for detractors of the barrier it was known as the McNamara Line.

The first proposed segment of the electronic battlefield—the Dual Blade system—was to be a ground barrier constructed with an assemblage of firebases, outposts, sensors, barbed wire, trip lines, and landmines that stretched inland from the South China Sea for about nineteen miles.[56] The second segment—designed to connect with the first—was hailed as futuristic at the time. This was a "virtual" air-supported barrier that cut across the Laotian segment of the Ho Chi Minh Trail, called Igloo White. The road network in this area would be seeded with myriad electronic sensors delivered by airplane. Bombers could then be directed toward the electronic signal, producing an automated link between sensor and shooter. McNamara jumped at the idea. Igloo White was operational by December 1967 and would serve in support of the Commando Hunt interdiction operations. The original McNamara Line proposed by the Jasons was associated with an unpopular secretary, and on March 19, 1969, the new secretary of defense, Melvin Laird, canned it. But the Igloo White segment of the project endured, running until the end of 1972 at a cost of around $1 billion per year.[57]

The Ho Chi Minh Trail was smothered with electronic sensors. Tens of thousands were parachuted in and camouflaged to mimic neighboring plants (the sensor field in Igloo White was usually composed of seven to eight hundred devices at any one time). Most sensors were acoustic or seismic, were battery operated, and lasted about thirty days. Each "listened" to truck and personnel movement and emitted a radio signal when triggered. Acoustic sensors were based on the navy's SONOBUOY, an early acoustic submarine-detection device. The ACOUBUOY was hung from tree canopies, whereas the SPIKEBUOY was lodged into the soil. The microphones on these sensors could be so effective that enemy conversations were overheard. The sounds of passing soldiers were amplified by seeding the ground with antipersonnel munitions. This included air-delivered gravel mines, tiny bomblets powerful enough to emit a bang when stepped on. Around 13 million were sewn every month, although the humidity of the jungle often extinguished their charge.[58]

Seismic sensors included the widely used ADSID (Air-Delivered Seismic Intrusion Detector), which was a spike-shaped device that detected vibrations in the ground. Finally, the ACOUSID was a dual seismic–acoustic sensor. Both acoustic and seismic sensors were joined by a range of other prosthetics: infrared devices that measured changes in heat; urine sniffers that detected ammonia molecules in the atmosphere; magnetic sensors that detected the presence of metal weapons; and low-light television cameras. Sensor warfare had arrived. Commanders could feel, hear, smell, and see the enemy from hundreds of miles away, moments before they were bombed. Major General Deane, head of the Igloo White project, testified that "the sensors denied the enemy his traditional cloaks of bad weather, jungle and darkness and detected his movements as he attempted to mount attacks."[59] The electronic sensor was "a solider with infinite courage," as one U.S. colonel proclaimed.[60]

Once a sensor detected a stimulus, it broadcasted a radio signal. In the air-supported system of Igloo White, this signal was usually heard by overhead airplanes. At the outset of the program, this aircraft was the air force's EC-121R, a four-engine transport with full crew. With around twenty crewmen on board, the air force looked to automate the sensor-relay process with drones. Between 1969 and 1971, modified commercial Beech Debonairs (QU-22s) were trialed on a limited basis. These were fully automated aircraft, although pilots did remain within the cabin. Despite never being fully rolled out, it was an early experiment in integrating drones in the electronic battlefield.

The airplanes retransmitted the radio signal to the brains of the Igloo White system, the Infiltration Surveillance Center at the Royal

Thai Air Force Base at Nakhon Phanom in Thailand. This nerve center, known as Task Force Alpha, was staffed by around four hundred air force personnel.[61] Due to the sheer volume of sensors, information relayed to the center was sorted by computer before being passed along to target analysts. "Data processing machines would furnish near or real-time receipt, processing, and display of sensor activities."[62] Two IBM 360-65 supercomputers digested and stored millions of bits of information. Banks of screens hung in the surveillance center's war room and displayed the road and sensor network of southern Laos. Activated sensors would light up once they passed threshold points. These illuminated lines on the map display became known as "worms." False alarms caused by animals (particularly frogs) or heavy rainfall were identified by algorithms calculated on the IBM computers.

If a positive target was produced, target assessment officers passed on their recommendations to an airborne battlefield command-and-control center, which was a modified AC-130 airplane. "Besides serving as an aerial traffic cop, the airborne battlefield command and control center used its radar, its computer that stored and spewed out target information, and its communications circuits to ensure that suitably armed aircraft attacked appropriate targets."[63] While a moving target could be hunted during the day, at night its pathway was extrapolated by onboard computers to produce a predicted static strike zone. This was the Commando Bolt technique—the splicing together of sensor-activation patterns and LORAN (long-range aid to navigation) radio coordinates. As Dickson writes, not only was the pilot blind to the battlefield below, but he would often "not even push the button that dropped the bombs—like so much else in Igloo White this was automated with the bombs released at the movement selected by the computer."[64]

It was this early automation of the sensor–shooter link that was truly groundbreaking about Igloo White. From beginning to end, the enemy existed as an electronic pattern of life. As Gibson writes, "At the point when Technowar reaches its technological apex, it turns completely into a representation. Indeed, the very name for a 'target' was '*target signature.*' And when the 'target' was destroyed, the lights on the screen went out."[65] Individuals became metaphors, worms that wriggled on a screen and then vanished.

But the success—and legacy—of Igloo White remains contested. In the cold light of the morning, nobody was sure who, or what, had been hit the night before. As one air force general admitted, "I don't think anyone can prove that we killed a single truck with the Commando Bolt operation."[66] After the North pushed past the demilitarized zone on March 30, 1972, resources were diverted away from Igloo White. The following year, the sensors fell silent, and there was no serious attempt to "Vietnamize" Igloo White. Nonetheless, this electronic battlefield laid the ground for a revolution in sensor surveillance.

The Electronic Battlefield, Part II: Bringing It Home

The 1970s saw Vietnam's electronic battlefield imported to the homeland. Much as the tear gas used in Vietnam was adopted by police in California, sensor technology used in Igloo White was redeployed by domestic law enforcement. The irony was not lost on everyone: the prospect of transferring the electronic battlefield to suburbia was terrifying. As Robert Barkan wrote, "While the ground War over there 'Vietnamizes,' the Nixon administration is quietly 'Americanizing' the war's technology, and the war on the home front escalates. The result: Americans, from marijuana smugglers to political dissidents to shopping housewives, are looking—though they may not know it—into the wrong end of the same surveillance devices that are spying on the Vietnamese"[67]

Under Operation Intercept, the antinarcotics search-and-seizure operations that ignited President Nixon's war on drugs, the U.S.–Mexico border would become a testing ground for the electronic battlefield. Sensors were deployed by border patrol in the summer of 1970, following a proposal for a surveillance system by Sylvania Electronic Systems (the same company would sell its sensors to homeowners and even the White House). As the *Washington Post* wrote, "Electronic sensors, used with varying degrees of success in tracing enemy troop movements in Vietnam, are being tested along the Mexican border to detect . . . narcotics smugglers."[68] Buried seismic sensors detected footsteps along the border and sent a signal to manned listening posts. Border patrol also began flying air force Pave Eagle unmanned remotely piloted aircraft—the same type of airplane used in Igloo White. The Ho Chi Minh Trail had arrived in Texas.

The post-Vietnam security market was big business. At an international conference in 1972, the convener stated, "There was a time when the public was very much upset about Big Brother. Now the public is beginning to accept this as a fact of life. They recognize, realize, appreciate and accept the fact that Big Brother is not really some alien being, but that he's their friend."[69] Not everyone agreed. A noted RAND Corporation (a military think tank) engineer, Paul Baran, argued giving law enforcement war gear was a mistake and may lead to one of "the most effective, oppressive, police states ever created."[70] U.S. policing was upgraded with Vietnam sensor technology. A range of night-vision devices and low-light and infrared cameras was rolled out by police forces across the country. As one army magazine wrote in 1972, "Night-vision technology developed by the U.S. Army to meet an urgent need of combat forces in Southeast Asia is . . . being applied increasingly to the peacetime 'service of man.'"[71]

There were more innocuous uses. Infrared cameras were used by the Environmental Protection Agency to trap polluters dumping waste in the Delaware River. As one newspaper reported in 1972, "The sophisticated equipment that spies on Communist troops moving along the Ho Chi Minh trail is doing some ecological snooping in the Delaware Valley."[72] Nonetheless, the Vietnam War was clearly a boon for domestic law enforcement. Indeed, the very first aircraft used by the FBI in the mid-1970s—Lockheed's near-silent Quiet Star airplane—was developed for battlefield observation during the Vietnam War. Sensor technology from Vietnam was also used to create the world's first electronic prison. The Federal Youth Center in Ashland, Kentucky, was a correctional institution that used an electronic detection system—PERIGUARD—developed by Westinghouse. The surveillance system involved ringing the prison with buried sensor hoses that detected any change in pressure. The limit of such a system was only in the imagination.

Take the proposal for the so-called Crime Deterrent Transponder System. This "futuristic" system was the brainchild of Joseph Meyer, an engineer who worked for the National Security Agency. In a 1971 paper, Meyer proposed fitting millions of American parolees, recidivists, and bailees with radio transponders. These transponders would broadcast their location and enable constant surveillance by the police. Instead of prison, criminals would be housed inside "an electronic surveillance and command-control system to make crime pointless."[73] The city would be converted into a gigantic sensor prison. "For urban areas, a mesh of transceivers would scan the streets, communicating with central computers to provide a public surveillance network."[74] Inside this electric sphere, a network of alarms, floodlights, and cameras would keep "subscribers" away from places like banks.

Such an artificial radio dome would engulf "the criminal with a kind of externalized conscience—an electronic substitute for the social conditioning, group pressures, and inner motivation which most of the society lives with."[75] If successful, subscribers would become law-abiding citizens that exhibited a passivity and routine that would make prediction of their movements inevitable: go to work, stay at home at night. Four decades after this proposal, police regularly use tracking devices on convicted criminals, sealing them inside an "externalized conscience." The electronic battlefield, in short, rapidly became a permanent condition of the modern city of technological civilization, as the battlefield and the homeland came to embrace each other in new and more complex forms of artificial enclosure.

The Electronic Battlefield, Part III: A Revolution in Military Affairs

In 1972 the Pentagon spent $219 million on electronic warfare programs (over $1.5 billion today).[76] The electronic battlefield was about to trigger a "revolution in military affairs." Malcolm R. Currie, the Pentagon's director of defense research and engineering, proclaimed before Congress, "A remarkable series of technical developments has brought us to the threshold of what I believe will become a true revolution in conventional warfare."[77] The networking of electronic sensors into the military's surveillance and communication systems was set to become a core practice.[78] Igloo White's direct link between target acquisition and the weapons system—the sensor-to-shooter link—heralded a much more dynamic form of targeting. Army general William DuPuy captured this when he stated, "What can be seen, can be hit. What can be hit, can be killed."[79]

This was the dawn not simply of the electronic battlefield but of the automated battlefield. In a 1975 article the *New York Times* described this automated battlefield as follows: "Wars fought by planes without pilots, between armies that may never see each other except as blips on an oscilloscope. Artillery able to hit moving tanks 10 miles away. Guns that select their own targets. Missiles that read maps. Self-operated torpedoes on the ocean floor. Laser cannons capable of knocking airplanes out of the sky. Satellite battles on the other side of the moon."[80] What was remarkable about this flurry of excitement was the acknowledgment that a global colossus had been unleashed. "Although Currie didn't say so," the *New York Times* continued, "he was describing a global, even universal, automated battlefield." The rise of the Predator Empire. As the article concluded, "Very soon, if all goes according to plan, it will be possible to think of the entire world as one big pinball machine."[81]

Indeed, a range of global military projects was in the pipeline in the early years of the 1970s, such as the navy's global sensor system SOSUS or the army's Remotely Monitored Battlefield Sensor System (REMBASS). These projects offered a glimpse of global surveillance, automatic killing, automatic everything. According to one Defense Advanced Research Projects Agency (DARPA) project officer, "Things just move too fast these days. We've got to get man out of the loop."[82] In a forward-looking 1974 *Newsweek* article, the journalist predicted that in the 1980s, "seated at his command console, the President will be able to order up a television overview of battlegrounds, to survey damage to both American and enemy cities and to evaluate potential target areas."[83] Here, of course, the *Newsweek* article correctly predicted the future involvement of U.S. presidents in the conduct of drone warfare, even if it was a couple decades too early.

Outside Vietnam the ongoing Cold War proved a hot laboratory for the electronic battlefield, and the Warsaw Pact threat in Eastern Europe was seen as a testing ground. As one industry article described in 1974, "The Army recognizes that success in any future European conflict will depend in large measure on mastery of the electromagnetic spectrum."[84] This became known as the "offset strategy"—the use of electronic warfare to balance the numerical supremacy of Soviet forces. In turn, Soviet military theorists began to recognize a military–technical revolution was taking place. They had been well aware of the smart bombs used by the U.S. Air Force in the Linebacker operations. Marshal Nikolai Ogarkov, the former chief of the Soviet General Staff, understood technology could act as a force multiplier if it was networked with sensors, computers, and precision weapons, forming what the Soviets would call a "reconnaissance-strike complex."[85]

In the United States post-Vietnam technological developments were institutionalized in the Office of Net Assessment (ONA). After working at the RAND Corporation since 1949, Andrew Marshall was appointed head of the ONA by Richard Nixon in 1973. Marshall, known by some as a "futurist-in-chief," would mentor a raft of Bush-era cabinet members, including Dick Cheney, Donald Rumsfeld, and Paul Wolfowitz.[86] What would be known as the "revolution in military affairs" (RMA) and later "defense transformation" captures the philosophy of the offset strategy. Indeed, many of the RMA's core tenets designed to lift the fog of war were central to McNamara's worldview. This vision is based on the networking of surveillance, airpower, and precision weaponry. Two decades after Vietnam, the Gulf War in 1991 would prove the site of a profound escalation in the electronic battlefield.

Saddam Hussein's Iraqi Army—the fourth largest at the time—invaded Kuwait on August 2, 1990. In response, the United States began amassing thousands of troops within Saudi Arabia under Operation Desert Shield, where they would remain for nearly six months. The UN Security Council approved sanctions, but President George H. W. Bush pushed for war. On January 12, 1991, the U.S. Senate passed a resolution authorizing force by the smallest margin since the War of 1812. Five days later, Desert Shield became Desert Storm, and the U.S. assault began. Airpower enthusiasts such as John Warden of the U.S. Air Force believed the whole operation could be completed from the air, with crippling strikes decapitating high-value strategic targets. Warden's ambitious and aggressive plan—Instant Thunder—was shelved, but its core faith in aerial bombing was adopted. The first night's strikes were unprecedented in the history of aerial warfare: seven hundred aircraft flew more than eight hundred sorties. It was almost a textbook performance of the U.S. military's AirLand Battle doctrine, the close coordination between the army and the air force.

The Gulf War was an information war built with a "mind-bogglingly complex communications system linking dozens of military and civilian satellites, more than two thousand personnel in fifty-nine communication centers, and tens of thousands of computers and phones."[87] As one government report concluded, "In future warfare, the struggle for information may play a central role, taking the place, perhaps, that the contest for geographical position has held in previous conflicts."[88] With unchallenged vertical domination, Bush brought the war to an end on February 28. In just six weeks, Desert Storm used more than double the laser-guided bombs released in North Vietnam over the nine months of Linebacker operations.[89] Despite precision-guided munitions accounting for less than 10 percent of total bombs dropped (17,000 precision-guided munitions were dropped compared with 210,000 unguided munitions), it was the television images of laser-guided bombs (LGBs) that framed the war.[90]

Marshall was impressed by what he saw in Iraq. The ONA began circulating its assessment of a military–technical revolution and later sponsored a series of RMA war games. The hypothesis was that RMA supersoldiers, "networked and supported by unmanned ground vehicles, UAVs, microrobots, and long-range precision fire," could achieve "massed effects" and overwhelm the enemy.[91] Admiral William Owens called this integrated approach the "systems-of-systems."[92] By merging sensors, computers, and precision weapons, the commander would obtain real-time Dominant Battle Space Knowledge to enable automated target recognition. Admiral Arthur Cebrowski was a kindred spirit. His coauthored 1998 paper "Network-Centric Warfare: Its Origin and Future"[93] is considered "the centerpiece of a whole new approach to war."[94] Central to both Owen's and Cebrowski's visions is the software that welds the war machine together: the algorithms, databases, and geographical information systems (GIS) that sift through information. NATO's 1999 air war against Serbian forces proved another paragon for RMA advocates after Slobodan Milošević surrendered without NATO deploying widescale ground troops.

Secretary of Defense Rumsfeld was keen to accelerate the RMA vision in the George W. Bush administration. He tapped Marshall to lead a defense review and appointed Arthur Cebrowski as head of the new Office of Defense Transformation. With these two hires, the RMA was no longer a pipe dream: it was an institutional fact. After the September 11, 2001, terrorist attacks, the idea of a horizontally organized netwar ignited.[95] President Bush, speaking after the fall of Kabul on November 14, 2001, stated, "Our commanders are gaining a real-time picture of the entire battlefield and are able to get targeting information from sensor to shooter almost instantly. . . . We're striking with greater effectiveness, at greater range, with fewer civilian casualties."[96] As war spread to Iraq in 2003, the U.S. Army began its modernization program, called the Future Combat Systems. As with the war games of the mid-1990s, the intention was to produce a fast and flexible infantry "built around a networked collection of sensors, drones, and weapons as well as manned and unmanned combat vehicles."[97]

The Iraq War started as a shock-and-awe replay of the Gulf War:

speed, simultaneity, and information dominance. Smart bombs, however, were used in 68 percent of strikes, compared with the Gulf War's 7 percent.[98] Targets could be hit in less than forty minutes after detection, compared with three days in the Gulf War.[99] Operation Iraqi Freedom also saw the widespread use of the GPS-enabled Blue Force Tracking system, which displayed the movement of friendly soldiers for improved situational awareness. By the time Bush addressed sailors on May 1, 2003, in the infamous Mission Accomplished speech, it looked as though technowar had finally produced a technovictory—kicking the Vietnam syndrome into the annals of history.

But there was a problem. Rather than repeating the dominance of Desert Storm in 1991, U.S. troops soon became bogged down in a low-tech insurgency, much like in Vietnam. According to Bacevich, "Rumsfeld's attempt to use the global war on terror as a device to validate his transformation agenda proved to be a massive miscalculation. Marrying the two together resulted in the undoing of both."[100] The descent into a protracted insurgency came on August 7, 2003, following the bombing of the Jordanian embassy. It was the first time insurgents had used a car bomb, and a week later, a suicide bomber attacked the UN headquarters in Baghdad. A downward spiral had begun, and the dream of a frictionless network-centric war was undone. Military strategy under General David Patreaus reappropriated the idea of winning hearts and minds, and General Stanley McChrystal would later install an industrial counterinsurgency killing machine. Indeed, one legacy of the Vietnam War—the electronic battlefield—was about to rub up against another: manhunting.

Manhunting: The Phoenix Program

Midway through the Vietnam War, the war managers implemented a new strategy focused on the South Vietnamese population. On July 3, 1968, General Abrams replaced Westmoreland as the new commander of the U.S military in Vietnam. His tenure crystallized a growing change in U.S. tactics and inaugurated a "One War" strategy aimed at securing South Vietnamese villages and eliminating the so-called NLF shadow government or Viet Cong Infrastructure (VCI). This widescale program of pacification was institutionalized in the 1967 Civil Operations and Revolutionary Development Support (CORDS) program. A joint military and civil program, it acted as an umbrella for the various pacification programs across South Vietnam. The aim of CORDS was to "win hearts and minds" through various social and rural programs, as well as dismantling the political apparatus of the NLF.

By this time, there was a range of different agencies in South Vietnam producing intelligence on the NLF shadow government. The CIA was thus eager to create a unified anti-infrastructure intelligence program, one centered around the practice of manhunting. The result was the infrastructure coordination and exploitation (ICEX) program. ICEX was signed into law by President Thieu in December 1967, who named it Phung Hoang, the All-Seeing Bird, which was translated to Phoenix. The aim of the project was to *neutralize*—that is, capture, convert, or kill—members of the VCI. As Douglas Valentine argues, Phoenix "was set up by Americans on American assumptions, in support of American policies."[101] Phoenix-style practices would be exhumed years later for the dirty wars of Latin America and the war on terror. As such, this bureaucratic machine of kill lists and manhunting teams is a crucial pillar of the Predator Empire.

Although the Phoenix Program was baptized under CORDS, its precedents stem from the counterterror (CT) teams recruited by the CIA in the late 1950s and early 1960s under the Mountain Scout Program. These squads—headed by Navy SEALs and CIA agents from the joint Special Operations Group (SOG)—were composed mainly of South Vietnamese and Chinese mercenaries who were anti-NLF. Around four thousand CT teams were eventually formed, divided into teams of a dozen or so members, operating across South Vietnam's forty-four provinces.[102] U.S. war managers believed individuals on the Phoenix backlists were part of a wider communist network. In reality, the so-called VCI was often constituted of generations of nationalists who had resisted colonial occupation. Other targeted individuals could be villagers caught in the wider war, pledging allegiance to revolutionary or government forces for safety—and fearing reprisal from both. "For most Americans in Vietnam," writes Stuart Herrington, "the dynamics of the Vietnamese villager's dilemma were impossible to grasp."[103]

The CIA's CT teams were later rebadged as Provincial Reconnaissance Units (PRUs), and became the main action arms of Phoenix alongside the South Vietnamese police. Indeed, prior to direct U.S. involvement, the Diem regime's security forces were already terrorizing and imprisoning tens of thousands of Vietnamese civilians. Of particular importance was the U.S.-supported Vietnamese Bureau of Investigation (VBI), which managed the central records depository and was the main arm for suppressing Diem's domestic opposition. The VBI's detailed records on the population were crucial. The 1962 Family Census included photographing and fingerprinting many families in South Vietnam, together with recording their political affiliations and other biographic information. With this biopolitical dragnet in place, the Phoenix program would later draw its power. In 1965 South Vietnamese and U.S. forces created an organization known as the Field Police. Over ten thousand Field Police officers entered rural areas to generate "public safety," an idea that stemmed from British counterinsurgency in Malaya.

Unable to comprehend that nationalism rather than communism was the main cause of local people supporting the NLF, counter-terror operations could be especially cruel. Instead of winning hearts and minds like CORDS, the CIA's strategy—pioneered, in part, by Saigon CIA station chief Peer De Silva—at times mirrored the political and psychological tactics of the NLF, creating "an instrument of civilian terror."[104] Indeed, "by the late 1950s, increasing numbers of American Special Forces were in South Vietnam, practicing the terrifying black art of psychological warfare."[105] William Colby, former CIA director, eventually told Congress that "a lot of things were done that should not have been done."[106] The second component of Phoenix was interrogation. In each South Vietnam province, a detention and interrogation center was constructed, called a Provincial Interrogation Center (PIC). These PICs were funded and supervised by the CIA.[107] They were reportedly the sites of horrific crimes carried out by South Vietnamese officials, including rape; gang rape; rape using eels, snakes, and hard objects; rape followed by murder; electrical shock; water boarding; dog mauling; and good, old-fashioned beatings.[108] According to the CIA's own estimates, anti-infrastructure operations neutralized around eighty thousand VCI.[109] Although numbers are contested, up to forty-six thousand people were killed under the program.[110] Tragically, as Jane Mayer reveals, "a Pentagon-contract study found that, between 1970 and 1971, ninety-seven per cent of the Vietcong targeted by the Phoenix Program were of negligible importance."[111]

The Phoenix Program, like Igloo White and CORDS, was heavily computerized. As Alfred McCoy writes, "From the 1960s onward . . . *computerized* information infrastructure emerged during the Vietnam

War with automated data processing and electronic communication."[112] This created a circular flow of information—an enclosed, electronic world in which computers were used to "centralize all data on the Communist underground, identifying cadres for interrogation or elimination by the agency's counterguerilla teams."[113] Beginning in 1968, the Viet Cong Infrastructure Information System—later rebadged the Phung Hoang Management Information System—collated population data from the Defense Intelligence Agency, the FBI, and the CIA, together with agencies across South Vietnam. According to Valentine, "At that point the era of the computerized blacklist began."[114] The belief in an automated counterinsurgency, much like automatic killing, was an epistemology shared by the war managers living inside the war dome. Killing by blacklist, in which war managers selected targets for assassination, would later become central to the Predator Empire's program of targeted killing during the war on terror.

Beyond Vietnam, Phoenix's methods were transferred to a set of training manuals for a range of Latin American governments under Project X.[115] Across Central and South America in the 1970s and 1980s, Project X materials were distributed to guide anticommunist forces in prisoner interrogation, aerial surveillance, wiretapping, assassination, and even the use of truth serum (sodium thiopental). As McCoy writes, "By the mid-1980s, counterguerilla operations in Columbia and Central America would thus bear an eerie but explicable resemblance to those in South Vietnam."[116] Phoenix-derived Project X methods, written into seven Spanish-speaking training manuals, were taught at the U.S. Army's School of Americas (later known as the Western Hemisphere Institute for Security Cooperation). Many of its tens of thousands of graduates have been implicated in atrocities across Latin America, leading to critics labeling it the School of Assassins.[117] Likewise, Project X methods were implicated in the enhanced-interrogation techniques pursued by the Bush administration during the war on terror. "Under the pressure of the occupation of Iraq in 2003, these brutal interrogation policies were revived and quickly proliferated to involve thousands of ordinary Iraqis."[118]

Project X, in short, left a winding document trail that connected black sites in Vietnam to black sites in Iraq, embodying the tactics for torture, imprisonment, and extrajudicial manhunting. The final legacy of Vietnam, the use of drones for widespread surveillance, further connects the Phoenix Program to the war on terror: only the predators on the ground would become Predators in the sky.

The Rise of the Drone, Part I: The Firebee

Drones played their part in the Vietnam War, where U.S. war managers integrated a variety of unmanned vehicles into the electronic battlefield. Several historical trends influenced the development of U.S. drone warfare before, during, and in the wake of the Vietnam War—particularly, an increased military desire for intelligence, surveillance, and reconnaissance (ISR). This desire for omniscience is, of course, as old as the military. Although it is difficult to break down the history of drone warfare into distinct phases, it is analytically useful to parcel out five key trajectories.

First, the drone was used as a practice target for military forces in the early twentieth century. Second, in the interwar period and into World War II, the drone was imagined as a kind of flying bomb that could be delivered behind enemy lines. Third, during the Cold War the drone was seen as a viable surveillance platform able to capture intelligence in denied areas. Fourth, the drone, since the war on terror, has been weaponized, fusing surveillance and killing, therefore becoming a combined hunter–killer, or a predator. The fifth stage sees the drone refashioned as a policing technology in domestic law enforcement. The evolution of these trajectories points to a much more diffuse—and everyday—use of drones for state power, one that mirrors the increasingly amorphous boundaries of the contemporary battlespace.

The idea of the drone covers a lot of ground. Although we often associate it with the military robots of today, drones, in some form or another, have been used for decades. One of the first recorded uses was by Austrians in July 1849 after they launched around two hundred pilotless balloons mounted with bombs against the city of Venice. Less than two decades later in the U.S. Civil War, Confederate and Union forces both flew balloons for reconnaissance missions. In 1896 Samuel P. Langley developed a range of steam-powered aerodromes, unpiloted aircraft that were flown successfully along the Potomac River near Washington, D.C. In those ninety-second flights, a glimpse of the future could be seen in the hovering aerodrome. The practice of aerial surveillance later emerged in the 1898 Spanish–American War when the U.S. military fitted a camera to a kite, producing one of the first aerial reconnaissance photographs.[119]

Of course, whenever we think of modern drones, we associate them with remote piloting. Remote piloting would not be possible without strides made in radio technology. Nikola Tesla demonstrated the first remote control of a vehicle at the end of the nineteenth century. On a pond in Madison Square Garden in 1898, the inventor and showman remotely controlled a boat, or what he called a "teleautomaton," with a radio signal. This was the first such application of radio waves in history, meaning that Tesla's U.S. Patent 613,809 was the birthplace of modern robotics. On that body of water floated enormous, largely unrecognized military potential.

The United Kingdom was an important hub for remotely piloted airplanes in the decades leading up to World War II. Captain Archibald M. Low of the Royal Flying Corps oversaw the construction of a number of remotely piloted airplanes fitted with explosives. This included the Aerial Target, which was launched in March 1917 from the rear of a truck in England. Although the lightweight wooden airplane failed to maintain its altitude, the Aerial Target did respond to radio control while airborne, thus launching Tesla's 1898 teleautomaton into the skies. The Larynx drone was a catapult-fired missile guided by autopilot that was tested in the English Channel and Iraq in the 1920s. Subsequently, the British-based Queen Bee (and later, Queen Wasp) was developed on a much larger scale and used for target practice.

Back in the United States, in 1917 Elmer Sperry, together with inventor and radio engineer Peter Hewitt, began construction of the radio-controlled Hewitt-Sperry Automatic Airplane, or "flying bomb." The Automatic Airplane was able to fly fifty miles while carrying a 300-pound bomb after being launched by catapult. Importantly, the pilotless airplane was stabilized with the addition of Sperry's gyroscope technology. The success of this project led the U.S. Army to commission a second project, the rail-launched Kettering Aerial Torpedo "Bug," developed by the Dayton-Wright Airplane Company. The Bug was essentially an aerial torpedo, pilotless and guided by preset controls.[120] In Germany a similar project was being pioneered by Wilhelm von Siemens between 1915 and 1918. The Siemens Torpedo Glider was a missile that could be dropped from a zeppelin and then guided toward its target by radio. The flying bomb, the Bug, and the Torpedo Glider were all early forerunners of contemporary cruise missiles. But the existence of such airplanes remained at an experimental stage.

World War II saw the large-scale development of strategic and tactical bombing. In the mid-1940s the GB-1 Glide Bomb was developed to bypass German air defenses. It was a workable glider fitted with a standard 1,000- or 2,000-pound bomb. Made with plywood wings and rudders and controlled by radio, the GB-1s were dropped from B-17s and then guided by bombardiers to their target below. In 1943, 108 GB-1s were dropped on Cologne, causing heavy damage. Later in the same war came the GB-4, or the Robin, which was the first television-guided weapon.[121] Although potentially revolutionary, the crude images broadcast by the drones could function only in the best atmospheric conditions.

The English-located project known as Operation Aphrodite was one of the more ambitious drone projects of World War II. The plan was to strike concealed German laboratories with American B-17 Flying Fortresses and B-24 bombers that had been stripped of their interiors and crammed with explosives. A manned crew would pilot these airplanes and parachute out once they crossed the English Channel. At that moment, a nearby mothership would take control, receiving live feed from an onboard television camera. Despite the inventiveness of the U.S. Air Force and Navy, Aphrodite was a military failure. It even claimed the life of Joseph Kennedy Jr. after his B-17 exploded over the English countryside. But the military was not about to give up: the development of Aphrodite, together with strides the Germans were making with the V-1 and the more sophisticated V-2 missiles, accelerated the development of U.S. unmanned projects.

After the war the special Pilotless Aircraft Branch of the U.S. Air Force was established to develop drones for training targets. Of the three drones developed, the airborne-launched Q-2 was the most important, becoming the father of a class of target drones built by the Ryan Aeronautical Company. The Firebee was first tested in 1951 at Holloman Air Force Base and would eventually become pivotal to U.S. surveillance in the following decade. The early Firebee could stay in flight for two hours and was capable of reaching heights of up to sixty thousand feet.

The Cold War backdrop influenced, even precipitated, the development of drones for surveillance. Although missiles and nuclear weapons were the most celebrated and feared technologies of this age, drones offered the U.S. military unprecedented visual access to denied areas. This benefit was reinforced by the perceived limitations—and dangers—of U-2 spy planes passing over the USSR. In 1960, for example, Francis Gary Powers was shot down over the Soviet Union while piloting a U-2 spy plane, an airplane created under the stewardship of the CIA.[122] As a result the Eisenhower administration scrambled to replace its manned reconnaissance program.

In 1960 Ryan Aeronautical Company (since acquired by Teledyne) proposed a version of its Firebee target drone, called Red Wagon, as a reconnaissance vehicle, which was later canceled. Eventually, however, in 1962, the year of the Cuban missile crisis, Ryan was given money from the U.S. Air Force's Big Safari research-and-development pot and built its first surveillance drones. These jet-powered Firebees went by several military designations: Ryan 147, AQM-34, and Lightning Bug. Launched from a Lockheed DC-130 Hercules airplane, the Lightning Bugs either flew preprogrammed routes or were controlled by airborne remote control officers onboard the Hercules. After performing their surveillance mission, the Lightning Bugs deployed their parachutes and were scooped up by helicopters under the guidance of drone recovery officers.

Lightning Bugs flown by U.S. Strategic Air Command were later used for surveillance in denied areas across an increasingly widening Cold War battlespace, including Cuba, North Korea, and the People's Republic of China. In November 1964 the *Washington Daily News* reported, "Communist China claimed to have shot down a U.S. reconnaissance plane with no pilot."[123] *Time* also reported, "Communist China held an official ceremony celebrating a 'major victory' in the shooting down of 'a pilotless, high-altitude reconnaissance military plane of U.S. imperialism' over Central-South China."[124] The U.S. military remained quiet about the wreckage from the secret project, much like it would decades later after the Iranians captured an advanced CIA drone—the Sentinel—in 2011.

Drones could also circumvent inclement weather conditions across Southeast Asia, which often hampered manned surveillance missions—especially during the monsoon season. This meant that U.S. pilots had to fly at increasingly high altitudes, since North Vietnamese aerial defenses prevented them from flying beneath the clouds. Low-flying drones had no such problems. Lightning Bugs were thus widely used across North Vietnam after Rolling Thunder ended in 1968, "relying on their speed and small size to elude heavy and effective North Vietnamese defenses."[125] These so-called Big Hunter operation missions were used extensively to surveil North Vietnamese sites at very low altitudes. Indeed, "often unknown to both those who looked at them and those that published them, many of the aerial views of North Vietnam that appeared in the American press were taken by the drones."[126] The jet-powered Ryan drones were also used for electronic listening missions (since they were able to intercept radio communications) and were even used as decoys to fool North Vietnamese defenses.

Between 1964 and 1975, Lightning Bugs flew over 3,500 combat sorties in Vietnam.[127] According to Colonel John Dale, then director of surveillance for Strategic Air Command's Fifteenth Air Force (SAC was the unit usually responsible for operating drones), the Lightning Bugs continually confounded MiG fighters over China, North Korea, and Vietnam by flying at these low altitudes. As Dale stated, "One of the things that is not said anywhere is, from October 1968 to November 1972—four years—we were the only aircraft flying in North Vietnam." He added, "[Because of CIA support] we were able to do so much for so long, because there weren't any politicians involved."[128] As one military magazine reported in 1970, "In Vietnam, the 147 drones [Lightning Bugs] were used so extensively and for such a variety of missions that Southeast Asia operations jocularly are referred to as the 'Tonkin Gulf Test Range.'"[129]

The Lightning Bugs were, however, expensive. In 1969 low-altitude drone operations were costing $250 million a year to maintain (over $1.6 billion today) and were funded mainly by the deep pockets of the National Reconnaissance Office's black budget.[130] By 1972 surveillance drones were fitted with LORAN technology, which drastically improved their reconnaissance capabilities. As mentioned, south of the seventeenth parallel drones were being trialed as electronic listening devices in Igloo White, although the Pave Eagle prototype was beset by too many problems. Nonetheless, as Thomas Ehrhard concludes, "The Lightning Bug drone was the most significant UAV operation in US history. . . . More than 1,000 drones were built with 200 lost in combat at a total cost of about $1.1 billion."[131] That translates to about $6.5 billion today.

Indeed, in addition to funding the Lightning Bug through its proxy, the National Reconnaissance Office, the CIA developed other drones during the Cold War, such as the Aquiline and Axillary airplanes. These prototypes were created in partnership with the Douglas Aircraft Company in the 1960s and tested at Area 51, although they were tabled in the early 1970s.[132] The Aquiline was a small, six-foot spy plane disguised to look like a buzzard and carried a range of sensors onboard, including a TV camera.

An emerging drone revolution was coming. The air force launched the Compass Cope program in the 1970s to increase the range and endurance of drones. This involved funding Boeing and Teledyne Ryan to develop high-altitude, long-endurance drones. Although ultimately unsuccessful, these prototypes were the most ambitious unmanned surveillance drones in air force history, capable of flying for over twenty-four hours while piloted from the ground.

At the same time that drones were getting bigger—more like the U-2s they were replacing—a range of minidrones were developed. Examples include the Praeire prototypes, which were capable of carrying laser designators and TV cameras. The air force also began experimenting with weaponized multimission Firebees. In May 1973 the Philco-Ford Corporation developed a laser designator that could be attached to a Ryan BGM-34B Firebee drone, with the aim of creating a strike drone. As one officer explained to the *Washington Post* in 1974, "That's where the big money is. You've got to be in on the kill."[133] In addition to Compass Cope, then, the air force began to consider strike drones by weaponizing the Lightning Bugs that had proved so successful in Vietnam.

The 1970s were accompanied by hyperbole about the end of the human pilot. Certainly, robot planes were getting smarter. As one Ryan publication from 1971 states, "The Firebee operator in 1951 was limited. . . . Today he can extend his eyes and hands and brain to the RPV without the presence of his body."[134] A year later, one pilot mused, "The day of automated warfare is closer than we think, with machines fighting machines."[135] According to an article in a 1971 *Air Force Magazine,* "For certain Air Force pilots, tomorrow's combat cockpit may be a swivel chair in a bomb-proof underground control center. From there, a USAF pilot may 'fly' by remote control . . . against targets hundreds of miles away."[136] Some of the excitement stemmed from the day the commander of the navy's Top Gun school was defeated by a Ryan drone in a 1971 war game.[137]

In the 1980s, although the U.S. Army maintained its drone program, government funding was cut. Reasons for this varied, including user reluctance in the air force (or what some would call a propilot bias), data transmission issues, and difficulties in retrieving drones (the Mid-Air Retrieval System [MARS] in Vietnam was cumbersome and costly).[138] Also, unlike the Lightning Bugs, which were funded by endless supplies of cash from secretive black budgets, research and development of these new drones were scrutinized by the public to a far greater extent. So the robotic torch passed to Israel in the 1980s. Israeli forces first used Pioneer drones in the early 1980s against Syrian forces, and today, Israel remains the world leader in drone exports. In any case, the Vietnam War was a remarkable period of U.S. drone development. Nearly all of the pieces were put in place for the Predator Empire—except, of course, the Predator.

The Rise of the Drone, Part II: The Predator

The drones so far discussed in this history have played two important roles: hunting and killing. As the war on terror unfolded, these roles would fuse together, with the Predator drone becoming a combined surveillance and weapons platform. The story of the Predator drone begins with Abraham Karem, the son of a Jewish merchant, who was born in Baghdad. His family moved to Israel in 1951, and by the 1970s, the young Karem was building aircraft for the Israeli Air Force. Karem built his first successful remotely piloted aircraft in 1973: a decoy to be used during the Yom Kippur War. The engineer lived and breathed aeronautics. Karem immigrated to Los Angeles in 1980. In a three-car garage, he began construction of a small remotely piloted aircraft. A year later, he wheeled out a bizarre cigar-shaped airplane called the Albatross.[139] This drone, patched together with plywood and fiberglass and developed by a handful of engineers, would change the face of U.S. military assassination.

Karem's Albatross proved its mettle. At Dugway Proving Ground in Utah, Karem demonstrated his drone could stay in the air fifty-six hours straight. This was a revelation. During the Vietnam War, U.S. Air Force drones could manage only a fraction of that flight time. The Aquila drone, for example, required thirty people to launch it, could fly for only a handful of minutes, and crashed on average every twenty flight hours.[140] Karem described this as "insanity itself." The success of the Albatross led to funding from DARPA, which sought a long-endurance drone more successful than the air force's Compass Cope. Karem built the Amber drone with this seed money—once again constructing it with a homemade team. The rapidly designed and assembled drone first flew in 1986. The Amber enjoyed some success, including demonstrating a flight endurance of forty hours and an altitude of close to thirty thousand feet by 1988. But it was insufficient for prolonged surveillance, since it was unable to carry large quantities of fuel or sensor equipment.

By the close of the decade, Congress was slashing the military's budget for drones. All research was consolidated into the Joint Program Office. This throttled aerial robotics and claimed Karem's company, which was eventually sold to the San Diego–based General Atomics in 1990. Karem stayed on and began work on a successor to the Amber: a runway-launched drone called the GNAT-750, which improved on the Amber in a number of ways. In 1993 the Pentagon issued a requirement to support UN peacekeeping forces in the former Yugoslavia. The Balkan conflicts saw increased importance given to aerial surveillance and dynamic targeting. To realize these tactical ambitions, long-range surveillance drones were needed, and the CIA was able to get its hands on drones fast, since the agency existed outside the congressional block on military drone research. James Woolsey was director of the CIA at the time and acquainted with Karem. According to Karem, "He asked his UAV programme officer to do the job, and they said it would cost $100m and take five years. He asked us, and we said it would be $5m and take three months. The rest is history."[141]

In 1994, under the codename LOFTY VIEW, the CIA operated the GNAT-750 in secrecy over Bosnia. The drones were stored in a hangar rented in Albania, in exchange for "two truckloads of wool blankets."[142] According to the CIA director, "I could sit in my office, call up a classified channel and in an early version of e-mail type messages to a guy in Albania asking him to zoom in on things."[143] But LOFTY VIEW's success was limited. The GNAT was vulnerable to inclement weather, and its radio could communicate only across a short distance. The range of the C-band line-of-sight data link was around 150 nautical miles. This meant the drone had to be controlled in relatively close proximity. The CIA tried to overcome this drawback by using intermediary aircraft to relay the radio signal, but the surveillance imagery simply had too far to travel: from a GNAT-750 to a relay aircraft to a ground station in Albania to a satellite circulating the planet to, finally, the CIA headquarters in Langley. An upgrade was needed. That upgraded drone was named the Predator, and it was fitted with a Ku-band satellite data link in its nose. No longer were operators restricted to the radio shadow of the previous GNAT model. Surveillance was revolutionized.

The Predator drone was first flown in the summer of 1994, where it was deployed to the Balkans under Nomad Vigil and Operation Deliberate Force in 1995 (the NATO air campaign against Bosnian

Serb forces). As the Predator proved its worth the air force became the lead service for operating the craft in Bosnia, establishing a UAV unit in 1995 called the Eleventh Reconnaissance Squadron at Indian Springs Auxiliary Airfield in Nevada (later Creech Air Force Base). Both the GNAT-750 and its offspring, the Predator, served simultaneously due to the massive demand placed on surveillance aircraft.

The Predator was again used during NATO and U.S. air strikes over Kosovo, where it was deployed—alongside the army's Hunter and the navy's Pioneer—to an unprecedented extent during Operation Allied Force in 1999. The Predator and Hunter logged around two thousand flight hours in Kosovo.[144] Although geographic intelligence was still limited by military software and interoperability issues, drones were incorporated directly into the military's kill chain—providing a more dynamic form of targeting. As General John Jumper explained in testimony, "We were able to successfully employ the Predator with laser only once before Allied Force ended, but in doing so, we developed a capability with great potential for rapid targeting."[145]

By the turn of the century, the Predator was only one of two operational drones in the U.S. military. In 2001, also at Indian Springs, a program under the Big Safari office successfully fitted an antitank Hellfire missile under the wings of the Predator. "I was not the guy who put missiles on the Predator," says Karem.[146] But the "dronefather" set the wheels in motion. The hunter was now a killer. In turn, as Mark Mazzetti writes, "the bar for war had been lowered, the remote-controlled age had begun, and the killer drones became an object of fascination inside the CIA."[147]

The Predator remains a surprisingly simple aircraft. It has a wingspan of fifty-five feet, a length of twenty-seven feet, a top speed of 135 MPH, and a flight ceiling of around twenty-five thousand feet. This makes it much slower than the Ryan Firebees that photographed Southeast Asia, but unlike the Vietnam-era drones, the Predator is able to loiter for much longer, beaming real-time satellite imagery to the other side of the planet. "What will stand out, what is historic for war, and human history in general," argues Peter Singer, "are the robotic weapons now playing greater roles on the battlefield."[148] The Predator drone, more than any other robotic weapon, is a concrete symbol of a form of electronic enclosure that annihilates distance through technology. The final chapter of this book will cover in more detail the next phase of drone warfare—stage five—which heralds an explicit policing role and a domestication of the electronic battlefield.

Summary: Atmospheric Warfare

The Vietnam War was a crucible for the Predator Empire. After nearly a decade of technowar, the electronic battlefield was established. Built with computer infrastructures, sensors, and the cybernetic systems that glued all the pieces together, the electronic battlefield installed a form of remote war that revolutionized the U.S. military. Parallel to this was a form of drone surveillance unprecedented in the history of flight and a manhunting operation that placed entire populations in the bureaucratic machinery of Phoenix. We can also add to this list of military power the ecological violence that wounded the landscape directly, scorching the earth with chemicals and bombs. How, then, do we unite all these seemingly diverse phenomena? Perhaps the idea of atmospheric warfare goes some way toward connecting the dots. The Vietnam War inaugurated a form of atmospheric warfare that sought to enclose the totality of life—both human and nonhuman—within the apparatuses of the electronic battlefield.

Atmospheric warfare is at once ecological and electronic, collapsing biological surfaces and radio signals inside the cybernetic spaces of the war machine. Military power is fundamentally invested in this process of atmospheric policing and destruction. In this way, atmospheres are always already technical, since assemblages of nonhuman actors constantly invade and destroy the spheres of human coexistence. Every hostile element of the environment is to be secured and pacified. By thinking volumetrically, then, we are required to articulate a form of geopower that encloses life in its human and nonhuman forms and surges across the technonatural husks of the planet. We can therefore think of atmospheric warfare as a *biopolitical project of enclosure* that captures, regulates, and at times eliminates the movement, form, and spaces of life. Atmospheric warfare brings, in short, the hybrid constellations of human and nonhuman coexistence into the interior worlds of the war machine.

In Vietnam this atmospheric form of security was manifested in the electronic battlefield. In 1971 Vietnam Veterans Against the War organized the Winter Soldier Investigation in Detroit. At the event over one hundred veterans gave public testimony to the crimes they had allegedly seen and participated in. Later that year, a second event was created to give voice to the pilots and weapons controllers of the war, who discussed the impersonal nature of the electronic battlefield. One veteran, Eric Herter, testified about the potential for atrocities greater than a "hundred My Lais." This new electronic battlefield inaugurated "the systematic destruction of thousands of innocent persons, of entire cultures by an automated electronic and mechanical death machine whose killing will be one-sided, unseen and universal. Those of us who testify are aware of the degree to which this hellish future is already upon us. We have seen the mechanical monster, the mindless devastation, the agony of simple people caught in the fire storm of our technological rampage."[149]

Decades after the Vietnam War, the electronic battlefield has grown in size and sophistication, realizing the prophecies of soldiers, generals, and journalists from the 1960s and 1970s. This mechanical monster now roams across the entire planet, a Leviathan let loose upon the world. The rush to secure the U.S. homeland in the war on terror has stretched the electronic battlefield across land, sea, and outer space. U.S. predators—both human and nonhuman—now surveil, police, and strike a globalized battlespace in a project of full spectrum dominance.

3

Full Spectrum Global Dominance

The Globalization of Violence

The electronic battlefield forged in Vietnam expanded beyond Southeast Asia after the conflict ended in the 1970s. This chapter explores the globalization of U.S. military control, paying particular attention to the infrastructures, strategies, and geographies that led to the development of the modern Predator Empire. Since at least the Cold War, the planet has been enclosed in a campaign of what the U.S. military calls full spectrum dominance: the control of terrestrial, maritime, atmospheric, and extraterrestrial spaces by a sophisticated war machine. As part of this surveillance, CIA and special forces teams regularly "find, fix, and finish" their prey wherever they surface on the global hunting grounds. While the electronic battlefield may have begun in Vietnam, during the George W. Bush administration its global power crystallized into the infrastructures of today. There is no outside or end for this Leviathan, only a titanic clash waged across land, sea, and sky.

In this new world order, the sanctity of the nation-state has been consistently challenged by the practice of targeted killings, which liquidate individuals with the press of a button. Where once assassins operated with cloaks and daggers, robotic killers now stalk the skies. The global lifeworld is hacked by drones, computers, and biometric systems alike, imprisoning populations inside overlapping electromagnetic enclosures. This chapter takes stock of the histories and geographies of U.S. military enclosure across the planet. Accordingly, a guiding inquiry is into how U.S. national security is embedded, or materialized, across the earth. The Predator Empire, as will become apparent, is entrenched within a worldwide military surveillance infrastructure.

Above all, deploying drones to survey and strike humans across the planet needs to be understood as a thoroughly global phenomenon: a brand of state violence that will soon be as recognizable as

Coca-Cola. Along with human migration, the circulation of capital, and the exchange of electronic data, state and nonstate violence takes place across increasingly complicated international borders. Indeed, many of the security issues endemic to technological civilization are international in scope yet lack easily definable territories: such as al-Qaeda affiliates metastasizing across Africa or the spread of Islamic State across Iraq and Syria. This global coimbrication provides the background to understanding the war on terror. Individualized small wars are creating something of an omnicrisis for the U.S. military, which has historically been geared toward fighting national armies. As such, the technologies and strategies of the Predator Empire have come to resemble a sophisticated electromagnetic manhunt.

The world is being remade into a battlespace. A boundless geographical logic has underwritten U.S. national security since the Cold War, especially since the terrorist attacks of September 11, 2001. There are multiple spaces of state power—terrestrial, maritime, extraterrestrial—that must, therefore, be explored. The U.S. military's practice of manhunting, so central to the war on terror, began with the CIA's hunt for Osama bin Laden, which weaponized the Predator drone and inaugurated a new regime of atmospheric power. Another crucial precursor is the historical act of air policing in Pakistan by the British Royal Air Force (RAF) in the 1920s. The spread of drone warfare, underpinned by what Chalmers Johnson calls an "empire of bases," has transformed the logic of U.S. military garrisoning across the planet. Critically, the dronification of state violence is driving a shift from a military Baseworld (with a large footprint and human presence) to a Droneworld (with a small footprint and greater robotic presence). The dronification of the U.S. Navy, which aims to create a fleet of automated robotic ships, is a further example of the expanding Predator Empire. The final frontier of military enclosure and, in many ways, the end game of full spectrum dominance is the occupation of outer space with spy satellites and other orbital weapons.

Manhunting

Targeted killing is not a clearly defined term in international law, although one United Nations rapporteur defines it as "the intentional, premeditated and deliberate use of lethal force, by States or their agents acting under color of law, or by an organized armed group in armed conflict, against a specific individual who is not in the physical custody of the perpetrator."[1] Crucially, what makes a targeted killing different from a B-52 carpet bombing of the Vietnamese jungle is that the attack is usually directed against an individual. The scale of the target is narrowed under what Gabriella Blum calls the "individualization of warfare." Individuals, often with no legitimate or internationally recognized status, are now able to enact forms of damage and destruction formally reserved for states.

If war and law once operated on the level of a collective (as with states and their armies), then today they are predicated on the rights and behaviors of individuals—regardless of national affiliation or territorial boundaries. As Blum writes, "Wartime regulation has evolved from a predominantly state-oriented set of obligations—which viewed war as an intercollective effort—to a more individual-focused regime."[2] The consequence of this, she argues, is that war comes to resemble a policing operation "in which people are expected to be treated according to their individual actions rather than as representatives of a collective."[3] In short, individuals are becoming (dangerous) agents of geopolitical power in ways that are historically novel while also finding themselves targets of state power.

While Blum believes this kind of evolution in the conduct of conflict contains a progressive kernel (one that promises to reduce collateral damage), it nonetheless legitimizes the modern practice of transnational manhunting. This individualized form of killing has been a key feature of the U.S. war on terror and has been facilitated by military drones, which as the United Nations observes, makes "it easier to kill targets, with fewer risks to the targeting State."[4] Indeed, by 2013, around 95 percent of the U.S. nonbattlefield targeted killings were conducted by drones.[5] Of course, targeted killings are not just the practice of the United States in its pursuit of the war on terror, nor are they carried out exclusively by drones. The key point here is the scale of the target is not a nation or its army: it is the individual body. Unlike Cold War strategic thinking, which obsessed over nuclear annihilation, modern U.S. war managers are fixated on erasing dangerous individuals from the surface of the earth.

The Israeli Defense Force (IDF) pioneered the use of drones in manhunting. In July 1971, following the so-called War of Attrition against Egypt, twelve U.S. Firebee drones arrived in the country to form Israel's first UAV military squadron. Fast-forward to the Second Intifada of 2000—the large-scale Palestinian uprising against Israeli occupation—and the unit held an active role in intelligence, surveillance, and security operations. During this conflict the IDF shifted its rules of engagement and popularized the policy of targeted killings.[6]

Less emotional than words like *assassination, targeted killing* has been used to describe military actions in repeated periods of violence between Israel and Palestine. Increasingly, the nature of this combat is characterized by its vertical dimension. As Eyal Weizman writes, Israel perfected what he calls "airborne thanatotactics," or the use of death as an explicit tactic to enforce security and political influence in the Palestinian areas, with Gaza becoming "the world's largest laboratory for airborne assassinations."[7] Although U.S. administrations have publicly protested Israeli assassinations, at the same time, as Weizman notes, different branches of the military have sought lessons from Israel in the art of robotic killing. Of course, despite the airborne thanatotactics waged by the IDF and its importation by the U.S. military, hunting humans is a decidedly old craft.

In one form or another, predation has saturated the blood of civilization. The prey simply switches as empires rise and fall: from hunting Native Americans in the so-called New World to hunting Africans in the pan-Atlantic slave trade. According to George Crawford, in a report published by the Joint Special Operations University, there have been at least fifty manhunts in American history, "from colonial efforts to pacify Native American leaders, through today's efforts to combat terrorists and insurgents."[8] He predicts the national security apparatus will begin to target individuals and networks rather than states, which means that constructing a "robust manhunting capability" has never been more important.

Given their centrality to state violence, past and present, manhunts cannot be understood simply as barbaric acts committed in a feast of primal rage but must be viewed as constitutive of the smooth functioning of sovereign power. The cruel war of all against all that defines Hobbes's pregovernment state of nature is not, therefore, vanquished by the arrival of civilization. As Grégoire Chamyou argues, "Contrary to what social contract theories claim, the state of nature is not anterior to the political order, it is not transcended by the latter but is, on the contrary, a virtuality within it that can always be reactivated in the mode of the exception to the law."[9]

Such a manhunt can take place over continents, and the idea of a national border is anathema to the Predator Empire. As Chamayou argues, "What is emerging is the idea of an invasive power based not so much on the rights of conquest as on the rights of pursuit: a right of universal intrusion or encroachment that would authorize charging after the prey wherever it found refuge, thereby trampling underfoot the principal of territorial integrity classically attached to state sovereignty."[10] The Predator Empire, in short, has sought to replicate this right of pursuit on a global scale, creating an unbounded hunting ground. The dronification of state violence transforms war into a diffuse campaign of extrajudicial predation. "The whole world, it is said, is a battlefield. But it would probably be more accurate to call it a hunting ground."[11]

The CIA Manhunt: "I Want His Head in a Box"

The CIA has dirtied its hands in the manhunting business since its inception. Given that the agency operates outside traditional military oversight, it forms a back channel for U.S. state violence that has been exploited consistently. In the 1950s, under the stewardship of Alan Dulles, the CIA rose to become a favored method of policing the planet. Under Operation PBSUCCESS, for example, the CIA supported the 1954 overthrow of democratically elected Guatemalan president Jacobo Árbenz. The coup set in motion decades of U.S.-backed military cabals responsible for sickening violations of human rights.

As mentioned in chapter 2, Phoenix-style training guides for Latin American governments were circulated by the U.S. military during the Cold War. One of these was an unsigned and undated manual called *A Study of Assassination*,[12] a document whose release was blocked for decades. "Assassination," as the macabre guide explains, "can seldom be employed with a clear conscience. Persons who are morally squeamish should not attempt it." Unsurprisingly, the document advises, in graphic detail, the best practices for murder. Nonetheless, by no means did a Phoenix mentality grip everyone in the U.S. government.

On December 22, 1974, journalist Seymour Hersh reported the CIA was conducting "a massive, illegal domestic intelligence operation during the Nixon administration against the antiwar movement."[13] As it transpired, since the 1950s the agency had been involved in a range of domestic black ops, which were detailed in reports that became known as the agency's "family jewels." These activities included the wiretapping of U.S. journalists, a string of illegal break-ins, the opening of U.S. mail, the surveillance of some 9,900 Americans, and assassination plots against foreign leaders such as Fidel Castro. Following these scandals, Senator Frank Church chaired a congressional committee to investigate the excesses of the U.S. intelligence community in 1975. The final report set in motion a variety of checks and balances on executive power, with the committee concluding targeted assassinations by the United States must be prohibited. It also required the CIA to brief Congress on its covert activities through what is called a "presidential finding." Senate and House Select Committees on Intelligence were also created to oversee U.S. intelligence activities. Assassination, however, would not disappear for long.

During the 1980s the CIA was overseeing the largest covert action in U.S. history: Operation Cyclone in Afghanistan. The CIA supported the Afghan mujahideen against Soviet forces throughout the decade. In 1983 President Ronald Reagan gushed, "To watch the courageous Afghan freedom fighters battle modern arsenals with simple hand-held weapons is an inspiration to those who love freedom."[14] Millions of dollars were sent to these freedom fighters to create a vast network of training camps, which a decade later, as Steve Coll observes, "would be referred to routinely in America as 'terrorist infrastructure.'"[15]

Two important events during the Reagan administration spurred the future of U.S. manhunting. First, in 1986 the CIA received an "an all-source intelligence fusion center for international terrorism." This was named the Counterterrorist Center (CTC), and it was tasked with coordinating a "clandestine service capability for preventing, pre-empting and/or disrupting international terrorist activity."[16] At the time a manhunt fusion center was controversial. Reagan's lawyers drew up secret memos defining the act of killing terrorists as self-defense, ensuring the CIA's hunting teams did not violate the 1976 assassination ban (just as lawyers for the Bush administration would in 2001).[17] Second, in 1987 the Pentagon's elite Special Operations Command (SOCOM) was formally established. The CIA's paramilitary wing, the Special Activities Division, was nurtured in synchrony with these highly trained soldiers.

The 1990s saw an institutional shift toward combating terrorism with a more individualized approach. Osama bin Laden appeared on the CIA's radar in the 1990s after funding the Afghan mujahideen. In January 1996 the CTC opened a new desk to track him down codenamed Alec Station. On January 23, 1995, Clinton signed Executive Order 12947. This piece of legislation approved the creation of a list of terrorists that could be individually targeted with economic sanctions (which later included Bin Laden). The Treasury's strategy was significant because it formally recognized the existence of specially designated terrorists and was therefore a key moment in the individualization of U.S. state power.

After two U.S. embassies in Africa were bombed in 1998, Clinton announced that Bin Laden had begun a terrorist war and approved the launch of seventy-five Tomahawk missiles at Afghanistan. In the summer Clinton signed off on the CIA's capture of al-Qaeda terrorists in Afghanistan. He would later approve the use of ambiguously defined lethal force.[18] Of course, this was controversial: by relaxing the ban on U.S. assassinations, Clinton had let the cat out of the bag. Manhunting was back. At the end of his administration, Clinton approved a Predator surveillance operation over southern and eastern Afghanistan to track Bin Laden. After spotting what the CIA believed to be al-Qaeda on September 7, 2000, a series of drone missile tests were completed over the next summer.

The newly elected Bush administration was cautious about weaponizing the Predator drone. After the attacks of September 11, 2001, however, most trepidation vanished. Afghanistan was to be invaded, and President Bush told his advisers he wanted the CIA to be first on the ground. On September 17, Bush signed a presidential finding that created a secret list of high-value targets whom the CIA was authorized to "hunt, capture, detain, or kill" without further presidential approval, "almost anywhere in the world."[19] This finding remains a key document in the globalization of U.S. state violence. Bush's presidential finding would encompass a suite of covert operations collectively known as GST, or Greystone, which included programs allowing secret prisons and enhanced interrogation. Of course, the 2001 Authorization for Use of Military Force gave the president the unique power to declare war against *individuals* and *organizations*. Cofer Black, head of the CTC, instructed the first CIA team that entered Afghanistan in 2001 on their mission: "Go find the al Qaeda and kill them. We're going to eliminate them. Get bin Laden, find him. I want his head in a box."[20]

Weaponized Predators reached Afghanistan on October 7, 2001.

President Bush was a fan. "Before the war, the Predator had skeptics, because it did not fit the old ways. Now it is clear, the military does not have enough unmanned vehicles. We're entering an era in which unmanned vehicles of all kinds will take on greater importance, in space, on land, in the air and at sea."[21] One of the CIA's first drone strikes took place on February 4, 2002, in Afghanistan's eastern province of Paktia. The agency unleashed a Hellfire missile at a "tall man" near the city of Khost, believing they had located the al-Qaeda leader. Despite denials at the time, the CIA had allegedly killed civilians gathering scrap metal. The site was Zhawar Kili, a mujahideen compound built with CIA support in the 1980s.[22] It was also the place at which Clinton launched Tomahawk missiles in 1998. Few targets on earth are scarred by the hubris of empire as much as Zhawar Kili.

But the precedent was set: Predator drones could execute the dull, dirty, and dangerous missions CIA operatives had performed since the Cold War. As the *New York Times* later reported, "The development of the armed Predator drone has made it much easier for the C.I.A. to pursue and kill terrorists in ways that would almost certainly not have been tried in the past for fear of the potential for American casualties."[23]

On July 22, 2002, Donald Rumsfeld issued a secret directive that instructed special forces to join the CIA's globalizing manhunt.[24] "Well a manhunt is certainly not what the armed forces of the United States are organized, trained and equipped to do," reflected the secretary of defense. "We may have to learn to do that and we are indeed learning to do it."[25] In a series of memos, Rumsfeld and his staff detailed a new vision for the global war on terror based on killing individuals outside traditional military channels. Manhunting would be lean, mean, and off the radar. Some CIA officers and Pentagon officials viewed the Phoenix Program as a model for the global war on terror.[26]

In November 2002 the CIA assassinated Qaed Salim Sinan al-Harethi in Yemen, an individual involved in the 2000 USS *Cole* terrorist attack in Aden harbor. This Predator strike was the first to occur *outside* a declared military battlefield. A U.S. citizen was killed in the strike, the first American to be droned by his own government.[27] At the time the White House defended the assassination, and Condoleezza Rice stated, "I can assure you that no constitutional questions are raised here," adding the strikes were part of "a new kind of war."[28] Others believed an important, even sacred, threshold had been crossed: the dawn of extrajudicial robotic warfare.

Vigilante Predators were now official U.S. foreign policy—and they were not going to be restricted to hunting within the confines of a declared battlefield. As the war on terror progressed, a Phoenix-style, dronified manhunt moved from the periphery to the center of U.S. national security. As Andrew Bacevich would later argue, "Much as COIN [counterinsurgency] supplanted 'shock and awe,' a broad-gauged program of targeted assassination has now displaced COIN as the prevailing expression of the American way of war."[29] Yet as the war on terror escalated it became clear the U.S. military's special forces could operate in a world even darker than the CIA's.

The World Is a Battlespace

"The ability of the CIA to move more swiftly than the military with just a fraction of the Pentagon's budget and manpower," writes Mark Mazzetti, "gnawed at Rumsfeld."[30] While the CIA was becoming more like the military, the military was in turn being engineered to resemble the CIA. Toward the end of 2001, Rumsfeld created clandestine special operations task forces with "blanket advance approval to kill or capture and, if possible, interrogate 'high value' targets."[31] Bush's presidential findings were expanded, "turning several nations in North Africa, the Middle East, and Asia into free-fire zones with regard to high-value targets."[32] In April 2002 Project Icon, or the Strategic Support Branch, was created by the secretary of defense to end his "near total dependence on the CIA,"[33] thereby wresting intelligence from Langley. The endgame was to remake the world into a single, unified battlespace populated by individualized, instantly accessible targets. As Jeremy Scahill argues, "Rumsfeld and Cheney were beginning to build up the infrastructure for waging an unaccountable war—and JSOC would be their prized weapon."[34]

For a long time the existence of the Joint Special Operations Command (JSOC) was a closely guarded secret. The unit was formed in 1980 in response to the botched mission to rescue fifty-two American hostages held in the U.S. embassy in Tehran. JSOC was created inside its parent organization, Special Operations Command (SOCOM), to ensure the United States had a specially trained counterterrorist force. Special forces personnel have since exploded throughout the war on terror. Fed by a multibillion-dollar budget, the number of shadow warriors has soared from thirty-three thousand at the start of the war on terror to more than seventy thousand today. By 2014 special forces—which includes Rangers, Green Berets, and Navy SEALs—had been deployed in 133 nations, capping a three-year period in which they had been active in over 150 countries on earth, conducting everything from training exercises to deadly night raids.[35]

Rumsfeld later signed an executive order on September 16, 2003, that established JSOC as the principal counterterrorist force of the United States and SOCOM as the lead combatant commander for Pentagon operations. This document contained a preauthorized list of fifteen countries in which counterterrorist action could be taken.[36] In the spring of 2004, Rumsfeld signed the Al Qaeda Network Execute Order (AQ ExOrd), which streamlined military operations across the world.[37] In effect, it formalized the freedom the CIA had enjoyed since September 2001. As if this wasn't enough, Rumsfeld pushed Bush to add language to the 2004 National Security Presidential Directive 38 that would cement SOCOM's global role in "finding," "fixing," and "finishing" terrorist suspects across the planet. Writing at the close of that year, Donald Rumsfeld penned a classified memo to his top advisers: "Today," he announced, "the entire world is the battlespace."[38]

Alongside JSOC, an elite intelligence branch of the special forces was created in 1981 called the Intelligence Support Activity, or the Activity. Capable of autonomous HUMINT (human intelligence) and SIGINT (signals intelligence) missions, the Activity is a common denominator of the war on terror. Its teams acquire various codenames depending on the task force with which they are deployed, including Gray Fox (in Afghanistan), Task Force Orange (in Somalia), and Task Force 121 (in Iraq and Afghanistan). The Activity "specialized in planting eavesdropping devices in hard-to-reach places—the devices could then link up to the large listening stations that the National Security Agency had set up around the globe."[39] These spying teams were deployed to "prepare the environment" in emerging target countries like Somalia, Yemen, Indonesia, Philippines, and Georgia.[40] In fact, not only did JSOC now have its own intelligence branch, but it would go on to acquire drones, reconnaissance planes, dedicated satellites, and even its own band of cyberwarriors.[41]

Only a small number of insiders were clued into JSOC's most clandestine task forces, known as Special Access Programs (SAPs).

These task forces fell under the umbrella program Operation Copper Green and were designed to extract intelligence from detainees, including the enhanced methods used against inmates at Iraq's Abu Ghraib prison. Significantly, the Bush administration defined SAPs as *clandestine* rather than *covert*. The difference is important. The CIA's covert activities fall under Title 50 of the U.S. Code, which states that foreign intelligence activities require a presidential finding. Clandestine operations, however, fall under Title 10 of the U.S. Code, which governs the armed services. These *do not* require congressional notification so long as they are "traditional military activities" that "prepare the battlespace" for future hostilities. The 1991 Intelligence Authorization Act largely exempted the military from reporting its activities to Congress if it was conducting secret operations it considered to be traditional military activities. As Mazzetti concludes, "The Pentagon could justify sending troops to any country in the world if it could make the case that the United States was at war inside that country—or might be at some point in the future."[42]

The Obama administration continued to prepare the battlespace around the planet. David Patreaus signed the associated Joint Unconventional Warfare Task Force (JUWTF) execute order in 2009, which continued to legitimate the U.S. military's clandestine manhunt. The shadow under which JSOC operates has spread far and wide. "We're the dark matter," said one U.S. Navy SEAL. "We're the force that orders the universe but can't be seen."[43]

The U.S. occupation of Iraq was a mecca for special forces. "Unlike the Green Berets," writes Scahill, "JSOC was not in the country to win any hearts and minds. Once JSOC took charge, the mission would no longer resemble anthropology. It was to be a manhunt, at times an assassination machine."[44] Consider the Bush administration's infamous deck of cards that symbolized the most-wanted targets who were subject to preemptive manhunting.[45] This program of assassination against Ba'ath Party members was designed to terrorize people into submission, which according to Douglas Valentine is "the point of Preemptive Manhunting, just as it was the point of the CIA's Phoenix Program."[46] As one former CIA counterterrorism chief stated in 2003, "They're clearly cooking up joint teams to do Phoenix-like things, just like they did in Vietnam."[47]

Under Stanley McChrystal's command, JSOC was catapulted to the front line of the high-value target manhunt. The sharp end of this knife was Task Force 121—a cadre of special forces kill teams (and CIA paramilitaries) with the authority to assassinate insurgents. JSOC operatives also ran a rendition program of black sites authorized under the covert Greystone program. So bad was the torture at one facility that even the CIA withdrew its interrogators.[48]

Afghanistan was not much different. Two wars were fought simultaneously: conventional fighting during the day and special forces raids at night. JSOC even maintained its own list of people to kill or capture. The Joint Prioritized Effects List (JPEL), as it was known, grew to over two thousand individuals.[49] Overall, there were five thousand special forces operating in Afghanistan after the surge of 2009.[50] John Nagl, a former counterinsurgency adviser to General Patreaus, described JSOC's campaign as "an almost industrial-scale counterterrorism killing machine."[51] Indeed, the number of covert raids in Afghanistan and Iraq soared from 675 in 2009 to 2,200 in 2011.[52] The manhunt doctrine had also spread to the regular U.S. military. In late 2006 marine commanders began taking hunting seriously and even consulted "a renowned African big game hunter."[53] As one colonel reflected, "We're hunting the enemy—those insurgents . . . hiding among the people. We're trophy hunting."

Much has changed since the CIA's 1954 guide to assassination in Guatemala. In the Predator drone, the CIA discovered a technology that allowed it to bypass all of the messiness associated with capture. In May 2004 the CIA's inspector general published a damning 106-page report on the agency's detention and interrogation program (the agency's grim atlas of extraordinary rendition was made possible by cooperation from at least fifty-four countries).[54] Following this, the CIA reportedly believed it would attract less criticism, as well as generate fewer legal headaches, if it began killing rather than imprisoning suspects.[55] The agency thus shifted from a jailer to an assassin in the space of a few short years. The drone was fundamental to this change. About 20 percent of CIA analysts had become targeters that hunted "for individuals to recruit, arrest or place in the crosshairs of a drone."[56] As Mazzetti concludes, "The Central Intelligence Agency has become a killing machine, an organization consumed with man hunting."[57] Together with JSOC, these agencies have driven a new approach to *preemptive planetary policing*. The chance of reversing this national security policy is unlikely. As Scahill observes, "Future US presidents—Republican or Democratic—will inherit a streamlined process for assassinating enemies of America, perceived or real."[58]

The Predators of the global war on terror thus roam across a planetary battlespace. By hunting individuals, the space of the target both *expands* to the scale of the planet and *contracts* to the level of the human body. "As a hunter state sees it, armed violence is no longer defined within the boundaries of a demarcated zone but simply by the presence of an enemy-prey who, so to speak, carries with it its own little mobile zone of hostility."[59] It is this restless double movement—a zooming in and a zooming out, both a localizing and a globalizing—that reformats the world as a battlespace. As Chamayou thus concludes, "In the logic of this security, based on the preventive elimination of dangerous individuals, 'warfare' takes the form of vast campaigns of extrajudicial executions. The names given to the drones—Predators (birds of prey) and Reapers (angels of death)—are certainly well-chosen."[60] Above all, this manhunt targets those individuals who threaten the smooth running of technological civilization.

Air Policing in Pakistan

No other place on earth has been subjected to CIA drone strikes with the same intensity as Pakistan's Federally Administered Tribal Areas (FATA). Between 2005 and 2015, there have been approximately 414 recorded drone strikes killing up to four thousand people, most of which have taken place in the agencies of North and South Waziristan.[61] JSOC forces have also conducted sporadic ground operations and drone strikes in this region. Indeed, sometimes special forces worked directly with the agency, as in 2011 when Navy SEALs were granted CIA authority during the execution of Osama bin Laden in Abbottabad (a process known as "sheep dipping").

No legal justifications were made by the White House when the CIA began its drone campaign in Pakistan. It wasn't until March 2010 that Harold Koh, then legal adviser to the Department of State, first argued the United States was in armed conflict against al-Qaeda and the campaign was consistent with its inherent right to self-defense under Article 51 of the UN Charter. But this justification, as well as others like it, has been widely criticized as an instance of "lawfare." In 2012 former U.S. attorney general Eric Holder argued a targeted killing was legal only if the "individual poses an imminent threat of violent attack against the United States." An imminent threat, as Holder also explained, did not require the definition of a "precise time, place, and manner." In other words, "imminent" was defined by whatever the administration said.

FATA is an international blowback capital, having endured external intervention since its creation. The territory's history is bound to its creation as a buffer region during British colonialism, when it was seen as a wild and ungovernable space.[62] In the Frontier Crimes Regulation (FCR) of 1901, the British codified a system of governance in which FATA became a semiautonomous state within a state beyond the regular rule of law. Upon independence in 1947, Pakistan adopted the FCR into its constitution. Therefore, many of the territorial complexities spun by the British lay beneath the human rights complexities of today. Furthermore, by the close of the 1980s, tens of thousands of Islamic militants (or "freedom fighters," as they were known back then), resourced by the CIA and other international organizations, were passing through secret training camps located in the tribal areas. After the Soviets retreated in 1989, the infrastructure, the fighters, and the skills remained in place like a ticking time bomb.

During their occupation the British military frequently battled against indigenous insurgencies in the mountainous Waziristan region. In a telling precedent, Royal Air Force (RAF) planes conducted aerial surveillance and bombardment of Pakistan's tribal areas in the interwar period, particularly the 1920s. At the time, this colonial practice was known as "air policing," and it sought to put down indigenous unrest across the British Empire, having first been trialed in Iraq. As Priya Satia writes, "Rather than rely on expensive and unpopular troop deployments, the British employed the fledgling RAF to patrol the country, coordinating information from intelligence agents on the ground to bombard subversive villages and tribes."[63] RAF wing commander R. H. Peck wrote, "Of all our mechanical improvements, and our new armoury of weapons, none has given us anything so great an advantage, and none is so admirably suited to warfare against wild men and in wild countries, as the aircraft."[64] In a 1924 secret directive, RAF chief of the air staff Hugh Trenchard wrote, "In warfare against savage tribes who do not conform to codes of civilized warfare[,] aerial bombardment is not necessarily limited in its methods or objectives by rules agreed upon international law."[65] For the British military, then, the result was a form of colonial policing without the costs of territorial occupation. "In the 1920s, air control thus served two related purposes: disciplinary surveillance and disciplinary punishment."[66] "Wild men" reappeared as targets decades later as air policing was reinvented by the CIA. Same place, different time.

On June 18, 2004, a Predator drone struck Nek Muhammad Wazir, a Pashtun tribesman from Waziristan and a veteran of the Afghan mujahideen. This attack was the first known CIA drone strike inside Pakistan. The assassination marked the beginning of a tortuous, even Kafkaesque, relationship between the CIA and Pakistan's Inter-Services Intelligence Directorate (ISI).[67] At least in the initial years of the relationship, Islamabad took responsibility for the killings, saying they had targeted an "enemy of the state." The Pakistan government was furious Nek Muhammed reneged on a peace deal, and so it allegedly granted the CIA airspace over FATA on the condition the Wazir commander be the first target. During the first four years of drone strikes, the CIA and ISI worked together in North Waziristan, sharing intelligence at an abandoned school in Miranshah and striking the tribal areas eleven times. This figure rapidly increased after the CIA switched to signature strikes, which no longer required targeted individuals be identified in advance of an attack. Between 2008 and 2012, there were over 340 drone strikes. As the former head of the counterterrorism center stated, "We are killing these sons of bitches faster than they can grow them now."[68] The escalation meant the ruse between the spy agencies became unstuck, and the Pakistan government began to publicly condemn U.S. drone strikes.

Despite surgery metaphors being used frequently to describe CIA drone strikes—particularly, targeted killings as being "precise"—the biopolitical damage done to the world must be reexamined, since it easily escapes the crosshairs of the military's framing. Drone strikes can cause widespread psychological trauma. As one Pakistani journalist explains, "If I am walking in the market, I have this fear that maybe the person walking next to me is going to be a target of the drone. If I'm shopping, I'm really careful and scared. If I'm standing on the road and there is a car parked next to me, I never know if that is going to be the target. . . . So, wherever we are, we have this fear of drones."[69] A single death creates ripple effects across an emotional landscape that cannot be digitized. Aerial assassination changes people's sleeping patterns, daily behavior, and friendship circles. Whether or not villagers are actually being monitored by Predators is irrelevant. The very *possibility* of surveillance is an oppressive form of atmospheric control that poisons everyday being-in-the-world. The biopolitical logic of drone strikes is not simply death, then: it is the ordering and policing of the lifeworld.

The value of situating the CIA's drone strikes in Pakistan alongside RAF bombing is in seeing both as a type of air policing that pacifies perceived threats to a civilized way of life. Moreover, as Satia argues, "Drones fundamentally alter a democracy's relationship to war; their purpose, like air control in the 1920s, is to make war virtually costless to Americans and thus to avoid public condemnation of the conflict itself."[70] The drone is thus a technology for policing and disciplining a population. As Mark Neocleous argues, drones are therefore capable of producing a totalizing form of air policing. "This is nothing less than a permanent police presence for the reproduction of order—air power as the *everywhere police*—in which the exercise of violence is an ever-present possibility."[71] Under such an understanding, drones become permanent mediators of technological civilization, nonhuman actors that enclose and police the population with an atmospheric power. Rather than end the war on terror, these technologies facilitate a forever war.

Targeting the Planet: The Digital Battlefield

Empires have conquered and colonized territory for millennia, but the Predator Empire does something different. As Alfred McCoy argues, this is a "nonterritorial American imperium."[72] It is an empire that seeks to dominate the electromagnetic spectrum and the patterns of life that flicker across its screens. Securing territory is no longer viewed with the same significance as securing the lifeworld, or what Foucault calls "living beings, and their environment, the milieu in which they live."[73] Here, Foucault invokes an ecological understanding of biopower, but the Predator Empire does not necessarily occupy these environments with physical enclosures. Instead, it remotely encloses these environments, placing them inside a twenty-first-century electronic battlefield, and thus produces a different kind of biopolitical space. Modern algorithms are crucial technologies for realizing this digital control in the war on terror.[74]

An algorithm is a software equation, or a set of sequential instructions, used for computer calculations. By processing vast sums of data, they can map and predict associations and probabilities. They are used in a national security context to locate "suspicious" or "dangerous" individuals within datasets (by analyzing, for example, travel itineraries, Internet browsing histories, credit card purchases, etc.). In addition to numerical data, Geographical Information System (GIS) programs can visualize and map individual movements across space and time. Geospatial intelligence (GEOINT) embodies this form of data collection and analysis. This intelligence activity requires, in turn, the widespread digitization of human activity across the globe. The organization responsible for delivering GEOINT to the U.S. intelligence community is the National Geospatial Agency (NGA). The NGA has a long background in cartography, mapping, and image analysis. Its organizational motto is Know the Earth . . . Show the Way . . . Understand the World. Headquartered at Fort Belvoir in Virginia, NGA analysts use GIS software to map, chart, visualize, and predict "what is happening at any location on the Earth" in order to gain "battlefield superiority."[75]

The use of computers to visualize the battlefield dates back to at least the 1990s. In the summer of 1994, the Army Science Board Summer Study Panel completed a study of the army's present and future technics. The panelists discussed the digitized battlefield embodied in the concept of Force XXI. This was the Pentagon's name for a series of experiments into digital war, with a number of generals touting it as the future of conflict. For the next few years, the army experimented with cutting-edge electronics. As a press release from 1997 stated, "Force XXI warfighters have satellite-to-laptop views of the battlefield. Enemy troops and vehicles appear as digital icons moving across computer screens."[76] During the NATO air campaign over Bosnia, the U.S. Air Force introduced a computer program known as PowerScene. This software package modeled the terrain of Bosnia so that pilots were able to fly virtually across the landscape. One of the most sophisticated GIS systems used by the military is GeoTime. It allows the user to map temporal data so that patterns over space and time can be discerned. This represents a transition from feature-based to activity-based intelligence.

Big data does, however, produce its own contradictions: hundreds of terabytes of information now swamp the U.S. intelligence community. As such, there has been a requisite demand to automate the sorting of data with algorithms. As Jeremy Crampton et al. write, "In GEOINT, the emphasis is on software that can integrate and parse the incoming data into actionable spatial intelligence by discriminating among billions of movements to identify those that are suspicious."[77] Automated tracking systems must be capable of finding and fixing suspicious spatial trajectories among billions of pieces of information. This process is known as "nodal analysis" and aims to bring to light dangerous patterns of life. Pattern-of-life analysis produces visual representations of the spatiotemporal pathway taken by an individual over the course of her or his daily movements. If the pattern of life is marked as a threat to national security, it can be eliminated under what's known as a signature strike. This forensic monitoring is what Derek Gregory calls a "militarized rhythmanalysis, even a weaponized time-geography."[78]

Suspicious patterns of life, such as a group of military-aged males meeting at a house or a convoy of trucks moving across a mountain road, may be enough to trigger a signature strike. CIA drone strikes in Pakistan have used pattern-of-life analysis to generate targets, mapping the space–time trajectories of entire populations. This new form of signature targeting is a defining feature of the Predator Empire.[79] Contemporary aerial policing is a preemptive, future-oriented immunitary conflict that hunts digital specters. The appearance of the category "al-Qaeda affiliate" in U.S. national security strategy marked the emergence of a far more epidemiological understanding of danger, in which the threat was located in what individuals *could potentially become*. The Predator Empire immunizes against not only actualized forms of danger, then, but also potential threats, those patterns of life that *may* become threats in the future. For Michael Dillon this means that "under the biopolitical regime of emergent emergency, it is no longer adequate to judge lifelike bodies in terms of the essence of that existential otherness definitive of the enemy alone, for every-body is a continuously emergent body-in-formation comprised of contingently adaptive rather than fixed properties."[80]

Predator drones have brought distant spaces into the gaze of pilots, targeters, and algorithms thousands of miles away. "Empire presents a superficial world," write Hardt and Negri, "the virtual center of which can be accessed immediately from any point across the surface."[81] Predator drones fold distant terrains by digitizing them, eliminating the geographical distance between the homeland and the battlefield. The direction of power is not simply an outward *projection* (as with historical examples of territorial conquest). Rather, imperial power today is one of *contraction*: enclosing faraway bodies inside the synthetic interiors of the war machine. Predator drones, connected by satellite and fiber-optic cables, produce electromagnetic wormholes across the planet that instantly connect faraway places and bypass the terrestrial obstacles of ancient empires. The Predator Empire thus marks the continuing evolution toward a topological empire in which space is unilaterally erased by technics.

The Skeleton of the Predator Empire

Baseworld

The Predator Empire's topological power still rests upon a global infrastructure of military bases and outposts. This infrastructure sustains what Johnson famously argues is "not an empire of colonies but an empire of bases."[82] These sites materialize a preemptive form of planetary policing. This constellation of bases forms the skeleton of the Predator Empire, providing the material infrastructure of targeted killing. U.S. base building reached its apex during the Cold War. As with the proxy wars of the time, the idea of garrisoning the planet was to contain communism. In order to ring the Soviet Union and China, a great chain of U.S. bases was constructed across Eurasia and the Pacific. In Japan alone, at the close of the Korean War, there were approximately six hundred U.S. installations.[83] After the Berlin Wall fell, instead of demobilizing this Cold War–era network of some 1,700 bases, "the United States imprudently committed itself to maintaining a global empire."[84]

Outside the Pacific, the Greater Middle East has been a crucial region for base building. As John Morrissey explains, "A proactive Department of Defense strategy for forging military links in the broader Persian Gulf region was vigorously pursued in the early 1980s."[85] Maintaining a forward presence in this oil-rich region remains a geopolitical and geoeconomic imperative. According to its latest official report, in 2014 the U.S. military maintained a global property portfolio of 562,000 facilities, located across 4,800 worldwide sites, covering a footprint of over 24.7 million acres. The number of active installations, including military bases, camps, and posts, is listed as 523, the plurality of which belong to the army.[86] "What is most fascinating and curious about the developing American form of empire, however, is that, in its modern phase, it is solely an empire of bases, not of territories, and these bases now encircle the earth in a way that, despite centuries-old dreams of global domination, would previously have been inconceivable."[87]

In terms of financial value, the 284,458 buildings the Department of Defense occupies around the world—covering a total of 2.2 billion square feet across forty countries—is estimated to be worth $570 billion. This inventory takes stock only of sites that exist in the "white world" of official, publicly known bases. The actual number of military bases is likely far higher than the 523 recognized officially. Nick Turse estimated that in 2011 the number of military bases was closer to one thousand, "an empire of bases so large and shadowy that no one—not even the Pentagon—really knows its full size and scope."[88] In 2015, David Vine counted eight hundred U.S. bases operating in about eighty different countries, costing in the region of $165 billion per year. As he writes, "Although few Americans realize it, the United States likely has more bases in foreign lands than any other people, nation, or empire in history."[89] There are multiple types of U.S. base: Main Operating Bases (MOBs), Forward Operating Bases (FOBs), Forward Operating Sites (FOSes), and Cooperative Security Locations (CSLs). Sometimes, the smaller FOSes and CSLs are referred to as "lily pads," since they allow the military to hop around the globe with a minimal footprint. In such cases, the Leviathan becomes bigger by going smaller.

Base building has been an integral component of the war on terror. At the height of the U.S. occupation, there were over five hundred military bases and outposts across Iraq.[90] During the JSOC manhunt, bases could function as extralegal spaces. Camp Nama (Nasty Ass Military Area), as it was known, was a prison used by special forces' Task Force 121 for intelligence gathering. Placards allegedly advised NO BLOOD, NO FOUL. As one Pentagon official said of the camp, "The reality is, there were no rules there."[91] Prisoners sometimes passed through Nama on their way to Abu Ghraib prison, the notorious jail where Saddam Hussein's enemies were held, tortured, and executed. In late 2003 revelations of U.S. prisoner abuse at Abu Ghraib emerged. By the time CBS broadcast photographs of humiliated detainees and the *New Yorker* carried an article written by Seymour Hersh,[92] tales of sadistic mistreatment were widespread. Such prisons in Iraq were also blowback sites, providing a recruiting ground for senior leaders of the so-called Islamic State.[93]

In the summer of 2004, Camp Nama closed its doors, and the JSOC unit moved to Balad Air Base, the largest of five American superbases in Iraq and home to around twenty-thousand air-conditioned troops. Balad's secure urban landscape came complete with a Subway, a Pizza Hut, a twenty-four-hour Burger King, a shopping mall, a hospital, a miniature-golf course, and even its own airline. "Like most towns, Balad has distinct neighborhoods."[94] Tucked away in one such neighborhood behind high walls was JSOC's new Joint Operations Center in Iraq, unofficially known as the Death Star. From this base JSOC ran its manhunt within the larger U.S. counterinsurgency. On the walls of the compound, banks of glowing screens known as Kill TV broadcast live imagery from overhead drones and night raids. As one journalist wrote, Balad was known as "the Death Star because of the sense that 'you could just reach with a finger, as it were, and eliminate somebody.' "[95]

In Afghanistan there were around 550 U.S.-led International Security Assistance Force (ISAF) bases at the height of occupation.[96] The true inventory of U.S. basing in Afghanistan is harder to count. As Turse explains, "When you add in ISAF checkpoints—those small baselets used to secure roads and villages—to the already bloated number of mega-bases, forward operating bases, combat outposts, and patrol bases, the number jumps to 750."[97] There were two mega U.S. installations in Afghanistan: Bagram Air Base and Kandahar Air Field. The former was the headquarters for special forces manhunts in Afghanistan. As with Iraq, JSOC ran its own detention program in Afghanistan, and prisoners were taken to a number of field detention sites or the Black Jail in Bagram. Like Camp Nama, these prisons were off-limits to the Red Cross and were reportedly the site of human rights abuses.[98]

Finally, the CIA's black sites drew much of Europe into a globalized network of detention and rendition. Some of the rendered individuals were taken to Guantánamo Bay. This naval base is both a concrete and a legal space synonymous with the excesses of the war on terror. It has been leased from Cuba since the 1898 Spanish–American War. The 1901 Platt Amendment guaranteed the United States would be able to rent coaling and naval stations indefinitely, leading to the perpetual lease of Guantánamo Bay. As Amy Kaplan explains, "The lease and attribution of limited sovereignty, which the Platt Amendment exemplifies, formed—and continues to form—an effective technology of imperial rule."[99] Furthermore, between 1902 and 1922 a series of important court cases was heard—the so-called Insular Cases—on whether those islands occupied by the United States were subject to American legal protections. In passing judgement, a new legal category was created, the unincorporated territory.

This was a space belonging to, but not a part of, the United States and whose people were neither citizens nor aliens. For Kaplan the legacy of the Insular Cases is thus a structural ambiguity that grants the U.S. government "great leeway in deciding whether, when, and which provisions of the Constitution may apply oversees."[100]

Although Guantánamo is certainly the jewel in the crown of the empire of bases, many U.S. bases in foreign countries exist as de facto spaces of legal exception. A Status of Force Agreement (SOFA) establishes the legal rights—and protections—of U.S. personnel operating on foreign soil. For Tom Engelhardt these indemnifying treaties represent "the modern equivalent of the colonial era's grant of extraterritoriality, so that American troops would be minimally subject to foreign courts or control."[101] U.S. troops in Okinawa, Japan, for example, have existed in their own legal bubble since the 1952 Japanese–American Security Treaty. This particular SOFA stipulates crimes committed by U.S. servicemen must first be handed to U.S. authorities, essentially creating a legal system within a legal system. Likewise, throughout the 1980s the Pentagon secured SOFAs from multiple Middle Eastern countries, regardless of their human rights records. Legalizing the foundations of the Baseworld is as important as laying its concrete. As Morrissey writes, "Today, the lawfare of the US military is a continuation of war by legal means."[102]

As U.S. military technology has become more sophisticated, the significance of mass, force, and territoriality has shrunk (although by no means disappeared). In particular, the pivotal role played by U.S. airpower has meant the importance placed on boots on the ground has waned. This recalibration from a labor- to a machine-intensive empire of bases has facilitated the growth of smaller bases. Developments in drone technology and communications mean that the military can now do more with less, which is why the days of gigantic MOBs may be overshadowed by a proliferation of smaller FOSes and CSLs (or lily pads). Projecting power is no longer a simple matter of building huge bases—it is about scattering U.S. surveillance and strike capabilities everywhere. A planet ringed with small but pervasive U.S. bases is crucial for delivering special forces or drones quickly, economically, and at times quietly. If, as Johnson argues, the twentieth century saw the U.S. military entrench a "globe-girdling Baseworld,"[103] then in the twenty-first century the Predator Empire seeks to install a global Droneworld.

Droneworld

Droneworld represents the evolution of Baseworld by robotic means. There are reportedly at least sixty worldwide military and CIA bases instrumental to the U.S. military's drone program.[104] To illustrate this transformation, consider Afghanistan once again. Operation Enduring Freedom was the longest war in the history of the United States. On December 28, 2014, the conflict was declared over by the U.S. government, leaving behind a skeletal force of around 10,800 troops (mainly special forces), which is considerably less than the 130,000 U.S. and NATO troops present in 2010. This reduction symbolizes the shift from counterinsurgency to counterterrorism. Despite this considerable reduction in human labor, a mixture of conventional air strikes and drone strikes continued. Indeed, as the humans exit the drones will stay in place to prolong the manhunt in Afghanistan. Counterterrorism fused with aerial policing appears to form part of what the U.S. military calls Operation Freedom's Sentinel. The choice of "sentinel" is appropriate, since it signals the U.S. military is watching or surveilling.

Afghanistan, argues Turse, will therefore become "a new hub for the American drone war in the Greater Middle East."[105] Of course, air strikes and drone strikes have been a constant part of U.S. military operations in Afghanistan. Despite CIA drone strikes attracting the most media attention, Afghanistan has always been at the heart of the unmanned wars: in 2012, for example, there were 333 drone strikes in Afghanistan, bringing the five-year running total to 1,015.[106] A pattern is certainly emerging, even if the phases overlap: the switch from a labor-intensive form of counterinsurgency to a robot-intensive counterterrorism.

After occupying Iraq since 2003, the U.S. military exited in 2011, leaving behind a legacy of brutal counterinsurgency and around seventeen thousand armed drone sorties.[107] Three years later, in June 2014, armed Predator drones began to assist a growing U.S. military and coalition operation against Islamic State militants across Iraq and Syria. The emergence of the Islamic State is partly a result of U.S. military blowback, although it cannot be simply reduced to that. As President Obama reflected in 2015, "ISIL [Islamic State in Iraq and the Levant] is a direct outgrowth of al-Qaeda in Iraq that grew out of our invasion. Which is an example of unintended consequences."[108] Since Obama authorized air strikes on August 7, 2014,

Islamic State strongholds across vast swathes of Iraq and Syria have been targeted by over ten thousand manned and unmanned coalition air strikes.[109] Crucially, it has been U.S. airpower rather than boots on the ground that has been the principal counterterrorist response to the Iraq War 3.0.

For the majority of the air attacks conducted in Iraq and Syria since 2014, the U.S. military has flown primarily from its gulf bases at Ali al Salem Air Base in Kuwait, al-Dhafra Air Base in the United Arab Emirates, and al-Udeid Air Base in Qatar.[110] The last is arguably the most strategically important base for air operations in the region. It hosts the U.S. Air Force Combined Air and Space Operations Center (CAOC), the $60 million nerve center of the U.S. military's operations across the Greater Middle East. It is the most advanced installation of its kind in military history, with miles of fiber-optic cable humming beneath the floor like a digital rhizome. This remote surveillance hub is, in many ways, a sophisticated descendant of the air force's command center, the Royal Thai Air Force Base, used during the electronic battlefield of the Vietnam War.

Judge Advocate Generals (JAGs)—or military lawyers—are on standby twenty-four hours a day for consultation on targeted killings. These lawyers ensure standard operating procedures are met, including the laws of armed conflict, the U.S. military's rules of engagement, and a set of specific instructions from a commanding officer. "Until the lawyers sign off on all three, the senior offensive duty officer cannot request permission from the joint air commander to fire weapons such as a Hellfire missile from a drone"[111] Given that most Predator and Reaper strikes in Iraq and Syria have been against dynamic rather than fixed targets, it means that al-Udeid analysts may be relying on sophisticated computer programs and SIGINT to make the case for or against eliminating threatening patterns of life.

Droneworld, as is often reported, is connected by satellites, linking pilot with drone from thousands of miles away. Since 2007, Creech Air Force Base has been home to the 432nd "Hunters" Wing, the first branch of the U.S. Air Force dedicated to flying drones. The base is also home to the Seventeenth Reconnaissance Squadron, which is the unit reportedly responsible for conducting CIA drone strikes in Pakistan.[112] Most communication is routed through undersea and underground fiber-optic cables. These carry 99 percent of international communications and connect many nations, which means the U.S. kill chain quite literally rests upon states that are *not* at war. Germany is an important example. Ramstein Air Base hosts the air force's Distributed Common Ground System (DCGS). This is the intelligence, surveillance, and reconnaissance (ISR) system for the U-2, Global Hawk, Reaper, and Predator. At present the DCGS is composed of around twenty-seven sites managed by the 480th Intelligence, Surveillance, and Reconnaissance Wing of the air force, which is located at Langley Air Force Base in Virginia.[113] The DCGS therefore acts as the nervous system for the U.S. global drone wars.

Yemen has been a target for drone strikes since 2002. After the CIA-backed Afghan mujahideen of the 1980s, a jihadist presence established itself within the country, led by future al-Qaeda leader Ayman al-Zawahiri of Egypt. For President Saleh these warriors could act as proxy fighters against southern secessionists and Houthi rebels in the north. So while al-Qaeda consolidated its territorial enclave in the country throughout the 1990s and after September 11, 2001, Saleh allegedly tolerated their presence. On February 3, 2006, former Osama bin Laden aide Nasir al-Wuhayshi and twenty-two others escaped from a maximum-security prison. Some of these convicts created al-Qaeda in the Arabian Peninsula (AQAP), a group Obama would later label "the most active and dangerous affiliate" of the terrorist group.[114]

U.S. counterterrorist operations against AQAP expanded in 2009 under JSOC's Special Access Program. On December 17 the U.S. covert war in Yemen began with a large aerial salvo. Controversially, JSOC reportedly used missiles packed with antipersonnel cluster bomblets at the al-Majala site.[115] Outrage spread across the country as forty-one civilian bodies were reported to have littered the wreckage.[116] But JSOC air strikes set a precedent that would only intensify as it stepped up its manhunt of U.S. citizen Anwar al-Awlaki. In June 2011 the CIA joined the hunt, starting its targeted killing program from a nearby drone base in Saudi Arabia. The targeters soon found what they were searching for. Under Operation Troy on September 30, 2011, al-Awlaki was killed in a CIA drone strike (on October 14 his son would be killed by a JSOC drone strike). Eric Holder, the former attorney general, later defended the practice of targeting Americans in a 2013 letter to the Senate. As he reasoned, "It is clear and logical that United States citizenship alone does not make such individuals immune from being targeted."[117]

In the summer of 2012, JSOC was given the green light to use signature strikes to hit suspicious patterns of life in Yemen.[118] At the end of 2013, one of the most deadly drone attacks since al-Majala struck a wedding procession, killing twelve and wounding at least fifteen.[119] Given the secrecy that surrounds the strikes, verifiable data is difficult to obtain, but it is likely that up to 639 people have been killed in over ninety drone strikes between 2002 and 2015.[120] But to what end? What kind of planet is Droneworld trying to secure?

Drone strikes have often enhanced the enemies they were seeking to degrade. General Stanley McChrystal, previous commander of JSOC, admitted, "The resentment created by American use of unmanned strikes . . . is much greater than the average American appreciates. They are hated on a visceral level, even by people who've never seen one or seen the effects of one."[121] While drones can *kill* threats, they can't *create* security. Yemen, at the time of writing, is experiencing some of its worst fighting in recent years, pushing the country to the brink of an all-out civil war involving religious factions and political parties from across the region. A stubborn fact therefore haunts Droneworld: victory is impossible with aerial counterterrorism. The focus of U.S. manhunting simply shifts as success in one area displaces—or creates—a problem elsewhere. Obama has, at times, acknowledged the open-endedness of the war on terror. In June 2013 he reflected, "What we can't do is think that we're just going to play whack-a-mole and send U.S. troops occupying various countries wherever these organizations pop up."[122]

But whack-a-mole is the only real option with Predators in the sky. Aerial assassination produces a forever war. Here, the logic of terrorizing a population, of eliminating dangerous patterns of life, takes precedence over any hope of winning hearts and minds. This is one of the central discontents of the Predator Empire. As with the technowar pursued in Vietnam, the very moment of tactical success—the production of death—produces a strategic failure. Perhaps this is the point. Jeff Faux argues, "Permanent empire means permanent war. But empires do not have to 'win' wars nor destroy all their enemies. There are too many of them. The strategy is to keep them at bay in order to demonstrate the empire's capacity to inflict murderous punishment on those on its periphery who challenge it, i.e., whack-a-mole."[123] Drone strikes endlessly reproduce the blowback that feeds an expanding Droneworld. As Chamayou writes, "The cull will be repeated periodically, in a pattern of infinite eradication."[124] We have, therefore, in the beating heart of Droneworld, a nihilistic loop of Predators clashing with prey.

The Dronification of Africa?

Billions of U.S. dollars have been pumped into Africa to build surveillance bases, arm and train allies, fight proxy wars, and conduct thousands of small-scale security exercises and missions. Yet Africa is not a new site of foreign militarization. In many ways it is still structured by old colonial legacies. The geopolitical situation, both then and now, remains highly complex. Nonetheless, the U.S. military has increased ISR, training, and counterterrorist missions across parts of West and East Africa, under legal authority from the 2001 presidential Authorization for Use of Military Force. The Obama administration has frequently insisted that al-Qaeda is like a tumor spreading around the planet—and since 2014, the same kind of vocabulary is used to describe the Islamic State. In 2012 General Carter Ham, the head of AFRICOM, stated, "Given the vast geographic space and diversity in threats, the command requires increased ISR assets to adequately address the security challenges on the continent."[125] The United States is not alone in its Africa pivot, of course: many African states are being courted by Chinese and Russian investment. Here, natural resources—rare metals, in particular—are a key geopolitical concern. Indeed, the geopolitical and geoeconomic are very much intertwined in Africa.

Since October 2008, AFRICOM, or Africa Command, has been a full spectrum combatant command. With around five to eight thousand personnel, it conducts operations alongside nearly every African military, averaging over one mission a day.[126] Headquartered in Stuttgart, Germany, AFRICOM coordinates training and security programs in Africa. For example, under the so-called Joint Combined Exchange Training program, units of special forces have trained foreign militaries in countries across the world. Between 2008 and 2013, the number of missions, exercises, operations, and other activities under AFRICOM's purview increased from 172 to 546. According to Turse, these training missions are designed to create U.S. proxy forces. "Like drones," he writes, "proxy warfare appears to offer an easy solution to complex problems."[127] In total, the U.S. conducted 674 military activities across Africa in 2014, nearly two missions per day. This represents close to a 300 percent jump in the number of annual operations, exercises, and training activities since AFRICOM was formed in 2008.[128]

Currently, there is only one "official" military base in Africa: Camp

Lemonnier in Djibouti. In support of Operation Enduring Freedom–Horn of Africa (OEF–HOA), Combined Joint Task Force–Horn of Africa has been charged with countering extremist organizations in East Africa and Yemen. After a period of renovations to Camp Lemonnier, a dilapidated former French colonial site, the base was fully functional by 2003. Personnel levels, which fluctuate, have numbered some four thousand troops, civilians, and contractors, with a smaller number of JSOC troops housed in a sealed-off compound. On average sixteen drones were taking off or landing at the Djibouti airfield every day in 2012.[129] In light of the attacks against U.S. diplomatic buildings in Benghazi the same year, there is now a 130-soldier crisis response unit (East Africa Response Force) stationed at Djibouti capable of rapidly responding to security crises across a 1,500-mile radius within eighteen hours.[130] As Craig Whitlock argued, "For the past decade, the Pentagon has labeled Lemonnier an 'expeditionary,' or temporary, camp. But it is now hardening into the U.S. military's first permanent drone war base."[131] Following a series of Predator crashes in 2012 at Camp Lemonnier, nearby Chabelley Airfield became the host of drone operations in the region. While targets in Yemen and Somalia were once central, the rise of the Islamic State has become a focus of drone activities under Operation Inherent Resolve. By October 2015, Chabelley drones had logged more than 24,000 hours of intelligence, surveillance, and reconnaissance.[132]

Since 2001 JSOC has overseen a covert war in Somalia, a country that hasn't had a functional government since 1991. The target of the manhunt includes members of al-Shabab (in Arabic, "the Youth"). At the outset of the war on terror, special forces and CIA operatives worked together to gather information on militants from an airfield in Kenya. Task Force Orange, as it was called, "ran a series of missions into Mogadishu to 'seed' the city with devices that monitored cellphone traffic."[133] These CIA and JSOC teams were monitored from Camp Lemonnier. JSOC began conventional air strikes against Somalia in January 2007, and by 2011 its fleet of Reaper drones began striking targets. That same year, the CIA opened up a compound in Mogadishu's Aden Adde International Airport where it would begin to train Somali intelligence officers.[134]

But why Somalia? In the wake of the U.S. invasion of Afghanistan, the Bush administration was concerned about al-Qaeda fighters resettling in the country. Under Operation Black Hawk, the CIA funded a coalition of warlords to track down the small number of al-

Qaeda operatives on CIA and JSOC kill lists. This meant U.S.-backed death squads were roaming the streets of Somalia, exacerbating what Scahill called "a course toward even further chaos and bloodshed."[135] As Somalia's security continued to disintegrate, small, regional Islamic courts began to rise, creating local justice systems based on Sharia law. These eventually united to become the Islamic Courts Union (ICU) in 2006. While the ICU had Taliban-style elements, the promise of security appealed to many Somalis. In response to the ICU's growing support, the CIA-funded warlords announced their own network, the Alliance for the Restoration of Peace and Counterterrorism, and declared war against the courts. Somalia then suffered some of the worst fighting in a decade.

Although al-Shabab began its life as the ICU's militant wing, it grew to forge its own identity over the years. Much of its domestic popularity would be garnered from the perception it was fighting foreign aggressors, even if the group's Wahhabi beliefs were at odds with the Sufism practiced by many Somalis. On June 5, 2006, ICU militia seized Mogadishu, routing the CIA-backed warlords. The U.S. military saw the victory as an Islamist takeover and was determined to topple the ICU. So Somalia's transitional government, supported by the Ethiopian army and U.S. forces, began a long war against the courts, beginning December 24, 2006. As Scahill writes, "It was a classic proxy war run by Washington and staffed by 40,000–50,000 troops from Somalia's widely despised neighbor."[136] As thousands of Ethiopian planes, tanks, and soldiers poured into the country in 2007, JSOC provided support from bases in Kenya, Ethiopia, and Djibouti under Task Force 88.[137] Throughout the spring of 2007, violence rocked Somalia, and JSOC intensified its manhunt amid the chaos. By the time Ethiopian forces withdrew in 2009, al-Shabab and al-Qaeda had seized vast swathes of Somalia. And by 2011 al-Shabab militia controlled more of the country than did the Transitional Federal Government.

The Pentagon responded, once again, by allegedly funding warlords and backing multinational proxy troops from the African Union Mission in Somalia (AMISON). Kenyan forces later invaded under Operation Linda Nchi, which lasted until the spring of 2012. Although complex, the years of warlordism and violence in Somalia are, at the very least, partly a result of U.S. intervention. On this Scahill is unequivocal: "Al Shabab's meteoric rise in Somalia, and the legacy of terror it wrought, was a direct response to a decade of disastrous

US policy, which had strengthened the very threat it was intended to crush."[138] Although weakened, al-Shabab has since been involved in a number of high-profile attacks on civilian targets outside Somalia. JSOC drone strikes, which have so far killed over one hundred people,[139] continue targeting al-Shabab commanders, "with drones now appearing to have superseded other, manned aircraft and cruise-missiles in the seven years since attacks began in Somalia."[140]

Outside Somalia the Sahel and Sahara regions have been the target of growing U.S. surveillance. Since 2005, Operation Enduring Freedom–Trans Sahara (OEF–TS) has been the name of the counterterrorist mission against al-Qaeda affiliates and other militants in the Sahara and Sahel region. At some point in 2013, OEF–TS was rebadged Operation Juniper Shield, although both names have been used interchangeably. In any case, through OEF–TS, AFRICOM trains partnering militaries, shares information and communications, conducts joint exercises, counters extremist ideology, and provides airlift and logistical support with ten African countries: Algeria, Burkina Faso, Chad, Mali, Mauritania, Morocco, Niger, Nigeria, Senegal, and Tunisia. Since 2007, under a program partly overseen by JSOC, around a dozen air bases have been enrolled in a growing pan-African surveillance network, using secluded hangars and isolated airstrips to conduct aerial surveillance.[141] Under the program Creek Sand, contractor planes have been flown in the hunt for al-Qaeda affiliates along the Maghreb-Sahel region since 2010.

Mali, a former French colony, has been a site of U.S. and French military intervention. The country has seen infighting for decades, with separatists from the north of the country waging an ongoing struggle for independence. This conflict was exacerbated in 2011 after arms and fighters from Libya's civil war entered the country. After the situation deteriorated, in 2013, France began a program of military assistance at the request of the Malian interim government. Operation Serval saw thousands of French troops spearhead a West African force to secure the country. These troops were supported by the U.S. military, with Predator drones providing actionable intelligence. Predators, as it turned out, had been deployed earlier that month in eastern Algeria, after militants took hundreds hostage at a natural-gas complex. As a result of ongoing drone surveillance in the region, in 2014 the U.S. military opened a second base in Niger, adding to a cluster of aircraft and drone bases in West Africa.

A similar picture can be seen of the Central African Republic. In the spring of 2013, despite years of U.S. assistance, its military was overthrown. France sent troops to its former colony to boost peacekeeping operations by the African-led International Support Mission in the Central African Republic (MISCA). Under Echo Casemate the United States offered African and French forces financial aid as well as airlift support and training missions. Staged from Burundi and Uganda, the operation contained a small number of U.S. personnel on the ground under Lion Forward Teams.[142] Since at least 2009, private planes have been used to hunt for Joseph Kony under the codename Tusker Sand. Airfields in Uganda and South Sudan have been used to monitor his Lord's Resistance Army across Central Africa.[143] This has involved surveillance flights over the Central African Republic, the Democratic Republic of Congo, and South Sudan.[144]

A pattern is clearly emerging in parts of Africa: U.S. drones, together with rapid-response special forces, are quickly becoming the de facto response to skirmishes, low-intensity conflicts, and counterterrorist operations across the continent. Within this new normal, countless innocent people are trapped in structures of violence that have persisted since European colonialism. And yet this kind of air policing appears here to stay. The material infrastructure—the roads, logistical networks, facilities, and drone bases—are spreading across the continent. In 2014 one AFRICOM captain suggested, "We have shifted from our original intent of being a more congenial combatant command to an actual war-fighting combatant command."[145] The threat posed by Islamic State and its franchises in Africa now complements al-Qaeda as a target of U.S. national security and is viewed similarly by war managers as an amorphous and *planetary* problem. At the close of 2015, Pentagon officials began to release plans for a new, integrated system of bases across Africa and the Middle East, mixing together larger "hubs"—such as military bases in Djibouti and Afghanistan—with smaller "spokes" in countries like Niger, where the United States already carries out drone operations. "The new approach would try to bring an ad hoc series of existing bases into one coherent system that would be able to confront regional threats from the Islamic State, Al Qaeda or other terrorist groups."[146] An architecture of hubs and spokes, of drones and special forces, aims at eradicating the tyranny of distance and brings the dangerous splinterlands of the continent under the watchful eyes of robots.

Ocean Power and the Rise of Robotic Waters

The technogeographies of Droneworld and their slow march across Africa and the Middle East are crucial to the contemporary manhunt. Yet the oceans and seas of the planet are fast becoming important spaces for the unmanned wars of the Predator Empire. The hydrosphere, as well as playing an important role in biological life, has always played an important role in geopolitical life. In the history of humanity, the ocean has consistently provided a medium for violence. From Viking raiders along the English and Scottish coasts to European colonialism in the New World, ships have been central to the entwined histories of empire, commerce, and slavery. In the near future the relationship between the ocean, the ship, and the drone will become more important. This is because ships and aircraft carriers—both miniature and mammoth—are able to produce movable spaces of U.S. sovereignty, thereby multiplying the reach of the Predator Empire from offshore.

Although ships have been a persistent feature of human exploration, the modern navy found its most sophisticated expression in the English Royal Navy. In the eighteenth century, the Royal Navy emerged as the unrivaled war machine of the seas, its dominance stretching back to the Tudor period. In 1495, just three years after Christopher Columbus arrived in the New World, King Henry VII oversaw construction of the world's first dry dock in Portsmouth. The Royal Navy's mastery of the seas reached its pinnacle after victory in the 1805 Battle of Trafalgar. Change was afoot, however, in the twentieth century as the role of ocean power slowly waned with the growing precedence given to airpower. After an economically bruising World War II, the British Empire scrapped all of its prewar ships. Then, it was the turn of the American fleet to ascend. The Continental Navy, the predecessor to the U.S. Navy, was first established in 1775. George Washington extolled the virtues of ocean power, proclaiming in 1781, "It follows then as certain as that night succeeds the day, that without a decisive naval force we can do nothing definitive, and with it, everything honorable and glorious."[147]

Within two centuries Washington's ambition had manifested. As of 2015, the U.S. Navy had 272 deployable battleships and over 3,700 aircraft comprising its carrier wing. The world's largest navy is staffed by 325,000 active personnel, 107,000 in reserve, and 195,000 civilian employees. Of particular importance are the U.S. Navy's carrier strike groups, which fuse airpower and ocean power. The *Nimitz*-class aircraft carrier is the centerpiece of these constellations. These leviathans are the largest warships in the world, capable of lasting for fifty years with just one midlife refueling. First deployed in 1975 (with the USS *Nimitz*), these $8.5 billion ships are powered by two nuclear reactors and can carry over sixty aircraft and five thousand crewmen.[148] At over 1,000 feet long and 252 feet wide, the vessels are capable of speeds over thirty knots. Such aircraft carriers are important to the U.S. Navy for a number of reasons, including symbolic shows of power projection in places like the Persian Gulf and the Pacific Ocean. But most significant, the carriers enable U.S. airpower to move in denied and remote areas of the world.

A number of *Nimitz*-class vessels supported U.S. and coalition troops during operations in Iraq and Afghanistan. More recent strikes against the Islamic State have used the USS *George H. W. Bush* carrier strike group, which floats in the Persian Gulf. Indeed, U.S. airpower, which once relied on ground-based infrastructures, has become radically *deterritorialized*. As the U.S. Navy itself boasts, "The carrier battle group, operating in international waters, does not need the permission of host countries for landing or overflight rights. Nor does it need to build or maintain bases in countries where our presence may cause political or other strains. Aircraft carriers are sovereign U.S. territory that steam anywhere in international waters—and most of the surface of the globe is water."[149] To put it bluntly, the vast majority of the earth's surface is a potential medium for unilateral U.S. operations.

Underpinning these bullish assertions is the idea of international waters, or high seas. This idea dates back to the early seventeenth century, when Dutch jurist Hugo Grotius first argued the seas could not be owned like the land. The waterworld is the property of nobody, or *res nullius*. This idea of *res nullius* caught on. For example, in President Woodrow Wilson's Fourteen Points—a speech delivered to Congress to justify intervention in World War I—the "freedom of the seas" was given as a crucial rationale for U.S. intervention. The freedom of the seas doctrine was formalized in the 1982 UN Convention on the Law of the Sea (UNCLOS) under Article 87(1), which states, "The high seas are open to all States, whether coastal or land-locked." The majority of the ocean is international water, which is defined as the space outside each country's exclusive economic zone (EEZ). The EEZ is an area that extends two hundred nautical miles from a nation's coast, going beyond the twelve nautical miles of territorial waters that surround a country. Conjoined to the idea of an open sea is the belief every ship is a sovereign jurisdiction. Article 92(1) states, "Ships shall sail under the flag of one State only and . . . shall be subject to its exclusive jurisdiction on the high seas."

Sovereignty, in other words, follows the ship: maritime vessels radically extend a sovereign's reach without the friction of terrain or international law. At times, this has been exploited during the war on terror. For example, a Somali terror suspect was detained for two months aboard a U.S. Navy ship in 2011. He was captured in international waters while traveling between Yemen and Somalia and was interrogated onboard the ship without charge and without a lawyer. Critics were quick to label the ship a "floating Guantánamo Bay," arguing, "There is some evidence that the US government is turning to detention at sea as a way of avoiding legal and political impediments in the treatment of terror suspects, both domestically and on the international stage."[150]

By operating from international waters, the U.S. military can move without status of forces agreements (SOFAs) or other forms of legal rationale. It is worth noting, however, that UNCLOS—which the United States has neither signed nor ratified—explicitly states that "the high seas shall be reserved for peaceful purposes." Nonetheless, in the case of the war on terror, terra firma has been frequently occupied, surveyed, and destroyed from the seas. A powerful navy and a network of island bases have enabled the U.S. military to occupy and move across a vast ocean territory. Aircraft carriers have added to this littoral network for decades, extending the U.S. formidable Baseworld to the waterworld. As Engelhardt argues, "For Washington, 'offshore' means the world's boundaryless waters and skies, but also, more metaphorically, it means being repositioned off the coast of national sovereignty and all its knotty problems. This change, on its way for years, will officially rebrand the planet as an American free-fire zone, unchaining Washington from the limits that national borders once imposed."[151]

There are lots of reasons why the U.S. return to the high seas is becoming an increasingly central strategy. The shift away from large-scale counterinsurgency in Iraq and Afghanistan to a geographically amorphous manhunt has globalized the hunting grounds and necessitated a flexible, agile force. Of equal significance is the proliferation of robotics and unmanned technologies. The enrollment of drones in the U.S. Navy is still at an experimental stage, but a trend is emerging: unmanned planes, unmanned ships, and unmanned submersibles are slowly transforming how the navy operates. In July 2013 the U.S. Navy's X-47B made history when it became the first drone to make an arrested landing, on the USS *George H. W. Bush.* This sophisticated drone represents a glimpse of the coming systems in which the U.S. Navy will invest heavily under the Unmanned Carrier Launched Surveillance and Strike (UCLASS) Program, to be completed by 2023.

This program puts the emphasis on computer-guided, semiautonomous drones that do not require direct piloting. "The objective of the UCLASS system is to enhance aircraft carrier/air wing operations by providing a responsive, world-wide presence via an organic, sea-based Unmanned Aerial System, with persistent intelligence, surveillance, reconnaissance, and targeting, and strike capabilities," according to a Department of Defense contract brief.[152] Unlike the current generation of fixed-wing drones, the MQ-8 Fire Scout is an unmanned helicopter that does not require a long runway to take off or land. This obviates the need for larger aircraft carriers. Indeed, the rush to construct an ocean Droneworld will not necessarily be made on the backs of gigantic *Nimitz*-style aircraft carriers. Smaller drones are revolutionizing the requirements for aircraft carriers. By going smaller, the U.S. Navy's surveillance capacities will be able to stretch further.

One future system is being built around an underwater mothership that communicates with semiautonomous or fully autonomous unmanned aircraft, ships, and submarines. The Defense Advanced Research Projects Agency (DARPA) is engineering a submersible called the Hydra, an unmanned underwater vehicle that launches modular drones from the ocean. As DARPA advertises, "Naval forces could deliver the Hydra system by ship, submarine or airplane to littoral ocean zones (shallow international waters near shorelines)."[153] The end goal is to create a distributed, global undersea system of drones that could be launched anywhere, anytime. The technology has an inherently defined global policing function. As DARPA's planning paper states, "The rising number of ungoverned states, piracy, and proliferation of sophisticated defenses severely stretches current resources and impacts the nation's ability to conduct special operations and contingency missions. The Hydra program represents a cost effective way to add undersea capacity that can be tailored to support each mission."[154] Of course, for decades—stretching all the way back to Operation Ivy Bells in the Cold War—navy submarines have infiltrated remote waters to eavesdrop on foreign governments.

The navy's trajectory toward smaller, more widely distributed drones is crystallized in a program called TERN, or Tactically Exploited Reconnaissance Node. Under this concept, developed by DARPA and the U.S. Navy's Office of Naval Research, drones will be able to survey distant geographies using extremely constricted runways on small ships. This would create persistent surveillance capabilities with a tiny, mobile footprint. As DARPA's program manager writes, "It's like having a falcon return to the arm of any person equipped to receive it."[155] If TERN's ambitions are realized, ground bases and aircraft carriers with long runways will become redundant. Smaller ships will not only be less vulnerable but also be able to enclose the seas far more effectively and rapidly. As DARPA advertises, "Effective 21st-century warfare requires the ability to conduct airborne intelligence, surveillance and reconnaissance (ISR) and strike mobile targets anywhere, around the clock."[156] Garrisoning the waterworld facilitates this form of global securitization.

This is not to suggest a single trajectory, however. The U.S. Navy still spends big bucks on big drones, such as the Triton MQ-4C drone, part of the Broad Area Maritime Surveillance Program developed by Northrop Grumman. These drones—which are based on the air force's RQ-4 Global Hawk—are able to survey close to three million square miles in a single mission using advanced radar and sensor systems. In any case, the U.S. Navy's carrier strike groups are set to be transformed, becoming the premier provider of lily pads that coordinate drone strikes and surveillance orbits across the globe. Currently, the drone's ability to open up exceptional spaces across the planet is restricted by ground bases and the complicated logistics that go with them, not to mention the complex legal SOFAs that must be signed. When the U.S. Navy fully exploits the fact that 98 percent of the world's land area rests within the flight orbit of TERN, the Predator Empire's reach will be boundless. While empires have traversed the seas for millennia, the watery enclosures of the Predator Empire have a distinctly robotic future.

The Final Frontier: Enclosing Outer Space

From soil to sea to outer space, the final frontier of the Predator Empire lies beyond the breathable atmosphere of human existence in the extraterrestrial orbits where only machines survive. Outer space, like the ocean, is part of the same full spectrum dominance strategy pursued by the U.S. military. We must, therefore, explore the satellites, missiles, and strategies that have foregrounded the growing entrenchment of a space war in the corridors of the Pentagon. The Global Positioning System (GPS) of satellite navigation is central to nearly everything the modern military does and has revolutionized drone warfare by enabling precision reconnaissance and targeting—remote split operations. Orbital space infrastructures, in short, enable the U.S. military to see and communicate across the planet.

The Cold War saw a protracted period of research into space technology by Soviet and American scientists. Sputnik, the Soviet satellite launched in 1957, was the first artificial Earth-orbiting satellite and propelled the Cold War space race. Sputnik was carried into space by the R-7 launch vehicle, the world's first intercontinental ballistic missile (ICBM). Over a decade prior to that, however, RAND (then part of the Douglas Aircraft Company) was investigating the launch of satellites in an important 1946 paper titled "Preliminary Design of an Experimental World Circling Spaceship."[157] Indeed, in many ways the roots of the looming space war actually date back to the V-2 rocket used by the Nazis, which influenced early missile designs in the United States and the USSR. The first American ICBM was the SM-65 Atlas missile, which began unsuccessful test flights in 1957.

As well as offensive capabilities, both the United States and the USSR began to construct defensive systems. This included antiballistic missiles (ABMs), together with ABM complexes that monitored incoming missiles. The development of these kinds of ABMs was restricted by the 1972 Anti-Ballistic Missile Treaty signed by President Nixon and Soviet general secretary Brezhnev. The agreement prohibited a nationwide missile defense system. It did, however, allow the United States to build the Safeguard ABM Complex in Grand Forks, North Dakota. In 1976—a year after the Safeguard site was constructed—the entire program was shutdown.[158] As it turned out, the system could be easily blinded if its radars were destroyed (a problem that hasn't really gone away).

The aerial bombardment of Serbia from March 24 to June 10, 1999, was one of the first major space-enabled wars, due to heavy reliance on satellites. For NATO this period was officially known as Operation Allied Force, and for the U.S. military, Operation Noble Anvil. While the Serbian military was ultimately subdued by precision U.S. airpower, Operation Noble Anvil revealed a paradox that still haunts the U.S. military today. As the Pentagon has become ever more reliant on space-based technologies, it has also become more vulnerable. Without the eyes and ears the satellites now provide, the military's high-tech systems would be blind and deaf. In spite of, or perhaps because of, this unwelcome paradox, the determination to militarize outer space has accelerated over the past two decades. As Johnson writes, "The United States now argues that it must totally dominate space to protect its new, casualty-free war-fighting technologies."[159]

On March 23, 1983, President Ronald Reagan delivered a national address that energized the antiballistic missile race and shattered the restraints of the previous administration. Reagan, who had campaigned to develop an ABM system, urged the United States to redouble its efforts. As he resolved, "I am directing a comprehensive and intensive effort to define a long-term research and development program to begin to achieve our ultimate goal of eliminating the threat posed by strategic nuclear missiles."[160] This ambitious project would become the Strategic Defense Initiative (SDI), overseen by the Strategic Defense Initiative Organization (SDIO). SDI was built on the idea of a planetary-wide defense shield that could intercept incoming Soviet ICBMs with ground-based missiles and orbital lasers. The proposal would be mocked as an unrealistic "Star Wars" fantasy and eventually collapsed. But this ostensible failure misses two important points. First, SDI paved the way for billions of dollars in defense spending and research. Second, it was the first step in militarizing space and enclosing the planet.

During the 1990s the conceits of cosmic power continued. In 1993 the SDIO was renamed the Ballistic Missile Defense Organization. Rather than engineering a global shield, the Clinton administration shifted its ABM strategy to focus on regional threats and rogue nations. Of course, the collapse of the Cold War should have ended the project, but there were too many vested interests. The Republican-controlled Congress accelerated antiballistic spending despite the enormous costs. Conservative defense hawks were convinced that the collapse of the USSR had everything to do with U.S. technological power and that throwing more money at missile technology "was a sure way to achieve perpetual domination of the world."[161] Missile lobbying came from the powerful right-wing think tank Center for Security Policy, which was funded by major weapons contractors and served as "the de facto center of the Star Wars lobby."[162] Frank Gaffney Jr. was the founder of this group and an important figure in the weaponization of the atmosphere during the 1990s. So too was Republican Curt Weldon, a board member of the Center for Security Policy. He obtained a resolution to create a congressional committee to assess the ballistic threat posed to the United States.[163] It was time to scare American lawmakers.

The Commission to Assess the Ballistic Missile Threat to the United States was chaired by Donald Rumsfeld.[164] Their 1998 report was heavily influenced by the military–industrial complex. Controversially, the Rumsfeld Commission, as it was known, contradicted a 1995 National Intelligence Estimate that a rogue nation would need ten to fifteen years to build a ballistic missile. The report warned rogue nations could acquire these capabilities within five years. "In essence, the Rumsfeld panel gave Star Wars boosters in Congress the quasi-official endorsement they needed to push the program forward."[165] By this point in time, the United States had already sunk $50 billion into antiballistic systems, none of which having proved workable.[166] The report's findings were leapt upon by proponents of a national missile defense shield, leading to the passage of the National Missile Defense Act of 1999. This called for the United States to "deploy as soon as is technologically possible an effective National Missile Defense system capable of defending the territory of the United States against limited ballistic missile attack."[167] President Clinton, however, deferred the deployment of the National Missile Defense (NMD) system. His reticence was unsurprising: an ICBM is incredibly difficult to stop effectively, given that the interceptor missile, or "exoatmospheric kill vehicle," is easily fooled.

After Bush came into office with Rumsfeld as his secretary of defense, Star Wars was back on the table. In December 2001 the administration announced it was withdrawing from the 1972 Anti-Ballistic Missile Treaty. On December 16, 2002, Bush signed National Security Presidential Directive 23, which instructed the Department of Defense to "deploy a set of initial missile defense capabilities beginning in 2004." The directive stated the missile shield must be *global* in scope: capable of defending the U.S. homeland, international forces, and even foreign allies, thereby eliminating "the artificial distinction between 'national' and 'theater' missile defense."[168] Reagan's planetary vision from 1983 was back. Bigger and better than before, the system would go by the name Ground-Based Midcourse

Defense (GMD). Since testing began in 1999, ground-based ICBM defenses in the United States have been unreliable. The GMD, which cost billions of dollars, is no exception.

In January 2001 Rumsfeld chaired the Commission to Assess United States National Security Space Management and Organization.[169] Unsurprisingly, the commission was influenced by the missile defense lobby. While the report acknowledged it was in the U.S. national interest to use space for peaceful purposes, it also recommended the construction of space defense systems. These would provide the president with "revolutionary methods" for space-based intelligence. Moreover, orbital defenses could protect the United States from a possible "Space Pearl Harbor." This incendiary phrase was repeatedly used to warn lawmakers that the military's heavy reliance on space technologies had rendered it vulnerable to enemy attacks. There are, after all, now approximately 1,200 satellites in space, of which over 500 belong to the United States. Of these, an unknown number serve as spy satellites for intercepting foreign communications and photographing the planet.

Space war is now, after two decades of lobbying, a strategic concern. No other country has antisatellite weapons in space, yet for U.S. galactic warriors the enclosure of space is viewed as an inevitable future for the military. Outer space presents the U.S. military with the same kind of opportunity as the seas—an environment devoid of direct sovereign control. There are no foreign governments to negotiate with and no need to adhere to legally binding SOFAs. "Best of all," writes Johnson, "the weaponizing of space enables [the United States] to project power anywhere in the world from secure bases of operation. It is, by definition, the global high ground."[170] One of the first documents laying out the U.S. military's ambition was *Vision for 2020*, published in 1997.[171] The booklet argues space is becoming the "fourth medium of warfare," after land, sea, and sky. As a consequence space must be controlled and, if necessary, denied to foreign governments.

The United States Strategic Command (USSTRATCOM) has overall responsibility for space operations. It was established in 1992 as the successor to the Cold War–era Strategic Air Command. Within STRATCOM there are two important organizations for coordinating activity in space. First is the Air Force Space Command (AFSC), which employs forty thousand personnel and operates over thirty-one military satellites. Within the AFSC lies the main hub for space-based operations, the Joint Functional Component Command for Space (JFCC Space). Known as Guardians of the High Frontier, JFCC Space operatives support U.S. ground forces across the planet. Perhaps unsurprising, the low Earth orbit (LEO) is congested with space junk—from spent rocket boosters to globs of frozen sewage. By its own estimates, the Air Force Space Surveillance Network (part of JFCC Space) tracks around twenty-three thousand objects every day (with hundreds of thousands of smaller pieces too small to track).[172] Given space debris travels at tremendous speeds, even minute fragments can severely damage and destroy satellites. This makes space war a disastrous proposition.

The U.S. 2010 National Space Policy affirmed its commitment to peaceful access to outer space, but it maintained that such "peaceful purposes" included using space for "national and homeland security activities." As the report states, "The United States will employ a variety of measures to help assure the use of space for all responsible parties, and, consistent with the inherent right of self-defense, deter others from interference and attack, defend our space systems and contribute to the defense of allied space systems, and, if deterrence fails, defeat efforts to attack them."[173]

In other words, the world's premier spacefaring nation affirms its status as the unilateral guardian of space, a galactic shepherd to the other nations that dwell in the upper and outer atmospheres. Similarly, the 2011 National Security Space Strategy argues, "Our military and intelligence capabilities must be prepared to 'fight through' a degraded environment and defeat attacks targeted at our space systems and supporting infrastructure. We must deny and defeat an adversary's ability to achieve its objectives."[174] Beginning in 2008, the United States established the multimillion-dollar Space Protection Program to coordinate the defense of U.S. space assets. This counterspace program includes funding for satellite-jamming technologies to disrupt adversary communications, although this is rarely vocalized in public.

The National Reconnaissance Office (NRO) is the U.S. intelligence community's eyes and ears in outer space (and, as chapter 2 details, was instrumental in the dronification of the Vietnam War). The NRO was formed by the Department of Defense in September 1961.[175] For much of its history, it was a classified agency, until its existence was publicly revealed in 1991. Around three thousand NRO personnel—drawn from the armed services, CIA, and civilian population—operate the U.S. fleet of spy satellites and deliver SIGINT and imagery intelligence (IMINT) to the other branches of the intelligence community. The NRO's black budget in 2013 was at least $10.3 billion, forming part of what Barton Gellman and Greg Miller call an "espionage empire."[176] Publicly, the NRO states it provides "innovative overhead reconnaissance" for U.S. national security. But as internal documents show, the NRO programs go beyond reconnaissance to collect and intercept global communications. As it states, "Space collection provides unique access to otherwise denied areas to provide persistent and responsive collection; and it does so without risk to human collectors or infringing upon the territorial sovereignty of other nations."[177] In short, this is another reworking of the high seas doctrine.

The NRO's first signals intelligence satellite was the 1960 GRAB, and its successor was POPPY, which intercepted Soviet radar communication until 1977. The first image intelligence satellite was the CORONA satellite, a system built in the late 1950s. Launched on August 18, 1960, CORONA's first successful mission photographed 1.65 million square miles of Soviet territory. Interestingly, in the days before imagery could be remotely transmitted, film had to be stored in capsules and dropped back down to the planet. Three thousand feet of film were captured in this way.[178] Indeed, CORONA was so successful it was used until 1972. For much of the Cold War, satellite intelligence from the NRO was used to record and estimate the number of missiles, planes, and submarines in the Soviet inventory. After the Cold War the NRO supported the 1991 Operation Desert Storm in Iraq. Since then, the NRO has played a pivotal role in the U.S. war on terror and continues to launch spy satellites.

The radio transmissions contemporary NRO spy satellites intercept must be downloaded back on the ground. Two of the biggest downlink facilities are located in the United Kingdom and Australia, part of the worldwide surveillance network called ECHELON, or Five Eyes. Menwith Hill, in the English countryside, is the largest electronic surveillance station in the world and is a cornerstone of the U.S. Ballistic Missile Defense program. Despite technically being an RAF base, the site is run by the National Security Agency (NSA). Around two thousand intelligence personnel and cryptanalysts, together with billion-dollar supercomputers, analyze the interpreted satellite data and transmit it to the NSA's headquarters at Fort Meade in Maryland.[179]

Pine Gap is the second-largest satellite downlink facility in the world. Built upon ancient Aboriginal land in Australia, the purpose of this remote base is to track the geolocation of radio signals and mobiles, pinpointing the whereabouts of enemy combatants in the Eastern Hemisphere. The facility, instrumental to the war on terror, was established by a 1966 Australia–U.S. treaty. Around one thousand personnel, mainly CIA, NSA, and NRO agents, control a set of geostationary satellites positioned above the Indian Ocean and Indonesia. The satellites are able to pinpoint the origin of a radio signal to within approximately ten meters. "Initially Pine Gap was collecting information—it was, if you like, listening in. It's now targeting weapons systems. It's also very much involved in the targeting of drones," explained former Australian prime minister Malcolm Fraser in 2014.[180]

From the launch of the first satellites, space has been a militarized domain, providing the eyes and ears for the U.S. military and intelligence community. The future points to not only more intense forms of militarization but also, potentially, more overt forms of weaponization—that is, the placement of orbital weapons in outer space. This has yet to happen, although both the United States and China have antisatellite capabilities. Some speculate the U.S. military's secretive X-37B unmanned orbital spacecraft is a space weapon of some sort, although its real purpose remains unclear.[181] The U.S. Congress has approved a big increase in space defense for 2016, totaling $5 billion over the next five years. Of course, the majority of UN states are against any form of space weaponization. To a large extent, UN legislation already prohibits space weapons, stemming from the foundational 1967 Outer Space Treaty. A big question mark therefore hangs over whether this kind of legislation will be able to hold back the emerging weaponization of space.

Summary: The Enclosure of the Planet

By the end of the nineteenth century, the United States had been violently settled. Railroads stretched from east to west, and a tidal wave of homesteaders, ranchers, gold miners, and families had migrated in one of the largest movements in human history. The frontier, the elusive beyond of American society, had been effectively closed. In the span of decades, not only had the Great Plains been settled, but so had the Pacific Coast. Of course, the closing of the frontier was really just another kind of mass enclosure, one that began centuries earlier in England. While the idea of the western frontier died with the enclosure of the American continent, numerous other frontiers have sprung up in the past decade, from Libya to Syria. To surveil these disparate areas and dangerous individuals, a globalizing infrastructure of air policing is being installed in the atmosphere. The enclosure of humanity's life-giving bubble with orbital weapons represents the most totalizing and daring imprisonment ever attempted.

From spy satellites circling the earth to Predators roaming over Mali to undersea drones swimming in coastal estuaries, the synergy between land, sea, air, and outer space is crucial to the full spectrum dominance of the planet. Alfred McCoy argues the war on terror has created a "robotic regime" preoccupied with dominating space. He writes the U.S. military is attempting to install a three-tiered "space shield" that envelops the earth within a totalizing surveillance network.[182] From satellites in the exosphere to space-based aircraft in the upper stratosphere to drones swarming in the lower stratosphere, the weaponization of the atmosphere is crucial to the future of the Predator Empire. As McCoy concludes, this regime "seeks to build a network of aerospace robotics, advanced cyberwarfare, and pervasive biometrics to envelop the earth in an electronic grid—allowing elimination of entire enemy battlefield formations through 'network-centric warfare' or incineration of a single insurgent with a drone-fired missile."[183] Such a pervasive space shield is the electronic battlefield of the twenty-first century—an atmospheric totalitarianism that poses innumerable threats to the human condition.

4

The Rule by Nobody

The Nonhuman Condition

Suffocating is one word to describe it. From the remotest of satellites to the most intimate of emails, it is difficult to escape the insidious creep of mass government surveillance today. The tools and technologies of state power keep exposing us, knocking down the walls of our private lives, all in the name of peace and security. But what kind of security? The interior spaces of technological civilization are inhabited by increasingly depressed, surplus, and alienated populations. In such times, when what is really secured is nothing other than a grinding inequality, we must uncover the deep links between the nonhuman infrastructures of state power and the human condition. Who are we, and what kind of world should we build on this splintering planet? Such philosophical reflection must be the start and end point for an examination of government surveillance. For Hannah Arendt it is *thinking*—and perhaps only thinking—that can begin to shake us from the dogmatic slumber that risks dragging us all toward a twenty-first-century totalitarianism.

The Predator Empire has already developed an extensive set of worldwide and domestic apparatuses for preempting potential terrorists by monitoring the communication of innocent citizens. An invasive form of surveillance has become a high priority for governments across technological civilization and is often viewed as a post-political issue. Yet the actual risk posed by terrorism is continually distorted by the media and readily consumed by society. Ulrich Beck argues such representations of terrorism, together with an obsession over anticipating them before they emerge, are "destroying the Western institutions of freedom and democracy."[1] Our insecurity, then, our seeming existential peril, is manufactured by a neurotic appetite (and industry) for fear, which leads to what Beck describes as a "totalitarianism of defence against threats."[2] Surveillance cameras, immigration restrictions, biometric control, bulk data interception, and a raft of other measures aim to immunize the population against global risk, forming a totalizing security enclosure. The paradox so often witnessed today is that under a spatially and temporally unbound "state of exception,"[3] the wholesale imprisonment of populations through electronic surveillance is no longer viewed as a risk to democracy but the very savior of it. It is the prison—or rather, the open prison—that provides the most stable, secure, and comfortable enclosure for the fearful publics of technological civilization to dwell inside. And these architectures of surveillance, in turn, affect us on a level far deeper than we perhaps dare to realize.

Arendt, whose writings on totalitarianism apply to the Predator Empire as much as they do to Nazism, writes that whatever force impacts us—artificial or natural—becomes part of the human condition. If the soil, the skies, and the seas constitute the ground for *biological* life, then the artificial things that populate the earth create the conditions for our specifically *human* lives. What Arendt calls the "human condition" can be understood as the *space* of human existence and activity, similar to what Peter Sloterdijk would later call a "sphere." This space is populated by artificial fabrications of all shapes and sizes: citadels of technology that sink their teeth into the ground and soar into the sky. Yet these creations bite back. They condition humanity with the force of a waterfall scouring the earth below it. No matter where we run or where we hide, we are always conditioned beings. As Arendt writes, "The objectivity of the world—its object- or thing-character—and the human condition supplement each other; because human existence is conditioned existence, it would be impossible without things."[4] Humans are creatures of the synthetic world: our anthropology is bound to technology.

Today, the central communication platform that mediates technological civilization is the Internet. This semiosphere fuses porn stars, terrorists, bankers, and cat videos inside an electric plasma. Trillions of e-mails are sent each year, and billions of devices are connected to the Internet. The Internet stems from the late 1960s ARPANET project (coordinated by DARPA), which connected research centers around the United States. Given its centrality for conditioning the human species and the world market, the Internet has rapidly become a domain to be privatized, policed, and militarized. No matter where we are, we can instantly connect to other people online. "Telecommunication," writes Sloterdijk, "is the rational faculty of haunting no matter what place in the world. All this contributes to a neutralization of space."[5] These topological systems present opportunities for state surveillance, criminal activity, and everything in between. U.S. targeted killings, for example, take place across this electric semiosphere. This is not a battlespace populated with tanks and soldiers but an electronic battlespace that endlessly records and regulates the minutiae of online and offline life. Space, the distance between us, is annihilated.

The Internet, furthermore, is a space targeted by the mass normalization of behavior, the binding together of consciousnesses. "Mass communication is even more effective than rapid transportation because it has the power of joining the nervous systems of inhabitants in a coherent space. It has the capacity to synchronize consciences in a very large semiosphere," writes Sloterdijk.[6] As such, digital enclosure can be defined as the appropriation, privatization, and policing of information that passes across the Internet. It is therefore part of the more general electromagnetic enclosure of humanity. The Internet has exploded to form a rhizomatic assemblage that taps directly into the biopolitical modulation of society. The Internet of today connects fridges with home security systems, tablet computers with wearable heart monitors, and drones with Amazon fulfillment centers. The Internet of Things describes the ubiquitous embedding of computing devices, or coded objects, into the existing Internet infrastructure.

Over the next several decades, billions of devices will be connected and woven into the flesh of technological civilization. The Internet of Things invites an unprecedented snooping of our dome-estic lives by corporations and the government: a home invasion from which we cannot possibly hide. As Catherine Crump and Matthew Harwood ask, "What will come next? Will eating habits collected by smart fridges be repackaged and sold to healthcare or insurance companies as predictors of obesity or other health problems—and so a reasonable basis for determining premiums?"[7] Binding these devices together is the idea of the always-online cloud. This integration of things inside an electronic, securitized semiosphere can be thought of as a type of digital enclosure: one that doesn't overthrow the material world as much as it modulates (that is, polices) the electronic grammar of human coexistence.

Since the Industrial Revolution, the economic intercourse of humans has been engineered by machines: from high-pressure steam engines in the nineteenth century to electronic computers in the twentieth century. Now, to an ever greater extent these machines have come to police our (international) social systems: surveying, digitizing, and eliminating other human beings. "We humans are becoming the dominant force for change on Earth," write Paul Crutzen and Christian Schwägerl, adding, "A long-held religious and philosophical idea—humans as the masters of planet Earth—has turned into a stark reality."[8] If, as Crutzen has popularized, we live in the Anthropocene, an age of human dominance over biological, chemical, and geological processes on earth, it is one not simply of human mastery but of machinic domination.

Moreover, the Anthropocene is an age of profound insecurity, paranoia, and exclusion, all expressed in the fervid construction of technological enclosures and a worldwide police force. A dramatic decline in existential security confronts all lifeworlds, even in the "secure" northern regions of the planet. These regions are dominated by surveillance Leviathans that science fiction writers of yesteryear could have only dreamed about. The key question to ask when faced with such a worldwide technological apparatus is, who, or what, is in control? Put another way, what happens when we delegate state power to technological systems? What happens when we offload and outsource the act of killing?

As discussed, the materiality of the world must be taken seriously when thinking about international relations, and the relationship between the nonhuman and the human is central to understanding the role of technology and machines in mediating, transforming, and executing (sometimes quite literally) contemporary U.S. state power in the age of the Predator Empire. Who has agency? Or what has agency? This is one of the most important debates in political geography and international relations, but it usually forecloses the idea that machines may hold sway over their human masters. As Nick Srnicek writes, "This is perhaps one of the most novel aspects of this new regime of technological governance: whether it is in the form of autonomous drones, algorithmic traders, automated surveillance techniques, or the automation controlling urban flows. The question that this all raises is 'what does *govern* mean when no decisions need to be made and where administration is automatic?'"[9] We must investigate, in other words, how state power is being augmented and mediated by technological systems.

Technologies embody certain momentums, inscribing a wider technogeography into the world. In this way, technologies become the material narratives that write our very being-in-the-world. This means that technology is unavoidably political, since it embodies, reflects, and refracts power relations. For example, Senator Frank Church, after glimpsing the machinery the National Security Agency (NSA) used to surveil the U.S. population back in the 1970s, was spooked by what he saw: "I know the capacity that is there to make tyranny total in America, and we must see to it that this agency and all agencies that possess this technology operate within the law and under proper supervision, so that we never cross over that abyss. That is the abyss from which there is no return."[10] Surveillance technology can enable a tyrannical form of government on a domestic and worldwide scale. As Glenn Greenwald writes, "When the United States is able to know everything that everyone is doing, saying, thinking, and planning—its own citizens, foreign populations, international corporations, other government leaders—its power over those factions is maximized."[11]

The role of technology in mediating, militarizing, and alienating the human condition is central. The human–machine nexus lies at the heart of contemporary U.S. surveillance and targeted killing. The Predator Empire, as will soon become clear, can be understood as a global immune system that secures the insecurities of living inside technological civilization. While the Predator Empire may attempt to immunize the U.S. homeland by surveilling and eliminating distant threats, it simultaneously secures the generic citizen of technological civilization. In this maneuver the Predator Empire defends and polices the vaunted American way of life, as it is expressed in the globalized economic logic of technological civilization. The conduct, logics, and rationale of U.S. mass surveillance—and the existential structures that underpin it—can be better understood via Arendt's ideas about the "rule by Nobody," as well as the philosophies of Michel Foucault and Bernard Stiegler. If surveillance risks interfering with or even extinguishing the very possibility of free thought, then we must investigate the histories, geographies, and programs of the most pervasive form of thought police ever created: the National Security Agency.

The Predator Empire: A Global Immune System

Biopolitics is a term used by Foucault to describe a new regime of government power that targets the life of the species, or what he terms "State control of the biological."[12] Unlike the sovereign power of kings and queens, who simply beheaded deviant individuals or else threw them in prison, biopower works its way into the intimate pores of the population's everyday lives. "This is a technology which aims to establish a sort of homeostasis, not by training individuals, but by achieving an overall equilibrium that protects the security of the whole from internal dangers."[13]

Biopower governs the life process, locking the species inside climate-controlled greenhouses, inspecting its vitals with electronic needles, and feeding it with optimal nutrients. Whatever deviates from the smooth survivalism of humanity is viewed as inherently threatening, and state apparatuses are put in action to control it. Foucault insists these security mechanisms must be installed to manage the random fluctuations inherent in a population of living beings. Life must be controlled. The question that follows is, how can human life—a seemingly rebellious force—be utterly implicated in its own imprisonment? From Joseph Stalin's Gulags to Adolf Hitler's concentration camps, the twentieth century witnessed *spectacular projects of enclosure* that controlled life by putting people to death. Protection of life, it seems, always reveals its hostile inverse. Self-preservation transforms the preservation of life into the pursuit of death.

The concept *immunity,* according to Roberto Esposito, explains why. Rather than see life and death as separate categories, he argues they are elements of the same whole. Whenever life is threatened, its very preservation inverts into a system of violence. "From this perspective, we can say that immunization is a negative [form] of the protection of life. It saves, insures, and preserves the organism, either individual or collective."[14] Immunity is the power to preserve life, and (geo)politics is nothing other than the construction of architectures for keeping life alive. In turn, since life is an unpredictable and emergent force, communities constantly build immunizing apparatuses across the biopolitical spectrum. While this process has a long history, Esposito argues the modern age has internalized an obsessive form of securitization. In his words, "All civilizations past and present faced (and in some way solved) the needs of their own immunization, but that it is only in the modern ones that immunization constitutes its most intimate essence."[15]

As discussed, Thomas Hobbes argues humanity in its raw state of nature is inherently violent. To shield humankind from this lawless and wretched landscape—a war of all against all—Hobbes argues for the establishment of a sovereign capable of ensuring peace. By protecting the body politic, the Hobbesian form of sovereignty becomes a type of immunity system that protects the life of the community by targeting those who would seek to destroy it. The history of civilization, read in this way, is the story of successive immunitary Leviathans, "in the sense that no society can exist without a defensive apparatus, as primitive as it is, that is capable of protecting itself."[16] Life and power collapse into each other as a singular process of self-preservation: immunity calls forth spaces of enclosure.

The Leviathan thus protects its commonwealth against the explosive power of life and thereby normalizes a biological disposition in the deepest recesses of the population, infiltrating the modern lifeworld with technics. If religion once immunized the masses, then with the long disenchantment of the world a significant theological baldachin was broken. This shattering of the ancient religious apparatuses "determines the need for a different defensive apparatus of the artificial sort that can protect a world that is constitutively exposed to risk."[17] The preservation of life through technics and other artificial prosthetics is a hallmark of modernity. Understood in this way, *enclosure is the production of spaces of immunity.* Yet what is immunized is not life as such (life in itself) but life only insofar as it is compatible with the twisted logics and mechanical contortions of technological civilization.

In other words, given humans share an equal capacity to kill each other in the Hobbesian state of nature, there can be little peace until this danger is checked. "Accordingly," as Esposito writes, "in order to save itself, life needs to step out from itself and constitute a transcendental point from which it receives orders and shelter. It is this interval or doubling of life with respect to itself that the move from nature to artifice is to be positioned."[18] The Leviathan is this transcendental configuration from which life is alienated so that it can be secured, and it is always already spatialized across a technological civilization bubbling with enclosed spheres. In the modern age men and women are individuated precisely through subtracting (that is, immunizing) themselves from the commons into their private, technically mediated comfort capsules. This subtraction defines the regime of private property. Both sovereignty and private property desocialize—or alienate—individuals from each other.

Immunity also strangles the meaning of liberty. Where once liberty was understood as a kind of positive, autonomous freedom, it now assumes its opposite meaning as enclosure. This negative meaning, which evacuates the ideals of communal living, defines liberty as an individual freedom *from* (a freedom *from* terror, a freedom *from* tyranny, a freedom *from* your neighbor). Modern liberty is the guarantee against interference from others, such that liberty is nothing other than security. The liberal subject of technological civilization wants nothing more than to be shielded from the masses. The defense of liberty in society, as expressed in the war on terror, inaugurates a corrosive manhunt against any individuals that threaten its existence. At every turn liberalism folds back on itself in a suicidal march toward totalitarianism. Individual self-preservation, the basis for political action in modernity, flips the preservation of life into the pursuit of death: the right to preserve life becomes the right to produce death.

The Predator Empire is just the latest protective apparatus of technological civilization, sworn to police a restive planetary body interconnected like never before. Above all, in today's global war on terror, it is *life*—in its dangerous, aleatory, and unpredictable nature—that is enclosed. As Jeffrey et al. argue, "If enclosure produces specific spatialities of inclusion and exclusion, these spatialities are also constituted by an *apparatus of biopolitical capture*."[19] Sometimes, the biopolitical apparatuses are more obvious than others, especially in the war on terror. In Afghanistan, for example, the U.S. military rolled out an extensive program of biometric capture—including a nationwide system for digitizing the fingerprints, retinas, facial images, and even DNA of the Afghan population by using handheld portable devices. As one U.S. military handbook describes, "Today, with biometrics on the battlefield, we can separate insurgents from the populace without moving anyone."[20] This counterinsurgency operation closely resembles the assembling of FBI databases in the U.S. homeland and can thus be understood as a type of battlefield forensics, which is defined as the scientific practices and technologies able to "uniquely identify, associate, and link people, places, things, intensions, activities, organizations, and events to each other in support of battlefield activities."[21]

U.S. drone warfare embodies and globalizes this immunitary defense system. Sometimes, the vernacular used by state officials confirms this paradigm exactly. John Brennan, head of the CIA, has appealed to an immunitary logic when discussing aerial assassination: "It's this surgical precision—the ability, with laser-like focus, to eliminate the cancerous tumor called an al-Qa'ida terrorist while limiting damage to the tissue around it—that makes this counter-terrorism tool so essential."[22] As it turns out, labeling terrorism as a cancer is a common narrative, with frequent allusions to metastasization made by the Obama administration. Former U.S. secretary of defense Leon Panetta said, "We have slowed the primary cancer—but we know the cancer has also metastasized to other parts of the global body."[23]

When Obama stated that "we will not apologize for our way of life, nor will we waver in its defense" in his inaugural address, he appealed to an immunitary biopolitics that remains the hallmark of the Predator Empire. The distinctiveness of *friend and foe* has transformed into the amorphousness of *patterns of life that threaten technological civilization.* To recall Michael Hardt and Antonio Negri, "Today it is increasingly difficult for the ideologues of the United States to name a single, unified enemy; rather, there seem to be minor and elusive enemies everywhere. The end of the crisis of modernity has given rise to a proliferation of minor and indefinite crises, or, as we prefer, to an omni-crisis."[24]

Immunization may protect and preserve, but by the same process it prevents any kind of transformation in the world. Those activities and groups deemed threatening are always increasing. Immunization has no limit. By protecting a community from the wolves of this world, the sheep begin to gnaw at their own flock.

An autoimmune disorder is a condition in which the immune system mistakenly attacks healthy tissue. We can see an autoimmune disease working today in the international system. The U.S. global immune system—the Predator Empire—is turning against the technological civilization that birthed it, attacking spaces of liberty and dissolving the barriers between war and peace, friend and enemy, homeland and battlefield. The conflict waged by Predator drones encloses the planet in a seamless battlespace and is unable—or simply unwilling—to differentiate between healthy tissue and the terrorist tumor the Obama administration has been so fond of surgically removing. As Esposito makes clear, "With the clear distinction between inside and outside weakened (and therefore also the distinction between war and peace that had characterized sovereign power for so long) sovereignty finds itself directly engaged with questions of life and death that no longer have to do with single areas, but with the world in all of its extensions."[25] The Predator Empire hacks, modifies, and modulates the global lifeworld, targeting dangerous individuals across a planetary population. There are few places to hide, and citizenship is no shield. Recall former U.S. attorney general Eric Holder, who defended the practice of targeting Americans with drones. As he reasoned, "It is clear and logical that United States citizenship alone does not make such individuals immune from being targeted."[26]

In its most paradoxical formulation, then, *technological civilization protects itself from itself*—from the contradictions that emerge out of its own apparatuses of enclosure. Planetary neoliberalism aims to achieve gigantic economies of scale by controlling and homogenizing human behavior, which in turn leads to intense civil unrest. The more successful technological civilization becomes at realizing its own efficiency (by automating the workplace, dronifying human activity, synchronizing consciences, robotizing love, life, and death), the more its denizens suffer. The war on terror is thus always a type of global *civil war,* an internecine conflict waged inside that great electric skin we call home.

Technological civilization, in short, must *immunize* itself against the very conditions it generates. Technics becomes *immunizing.* As Stiegler writes, "Technicity, as a system, constitutes the artificial and social system of predation and defense from the beginning of humanity."[27] Unless this form of totalizing power is checked—the likes of which we have never before seen—a pathological totalitarianism is surely on the horizon. Technics is not going anywhere, and the tendencies it harbors are treacherous to those who would seek to master it democratically. Taking it to its most radical conclusion, Arendt writes, "In short, the seemingly irresistible proliferation of techniques and machines, far from only threatening certain classes with unemployment, menaces the existence of whole nations and conceivably of all mankind."[28]

The Predator Empire thus names the global immune system scrambling to control the disaffected societies across the planet. We may wish to imagine that remote assassination takes place only "out there" in the darkest borderlands of the planet. We may comfort ourselves that NSA wiretaps only target the bad guys. But in reality these infrastructures infect the human condition on a far deeper, more insidious level. Such is the existential emergency with which we are confronted today.

Kafkaesque Killing

"Someone must have been spreading slander about Josef K., for one morning he was arrested, though he had done nothing wrong."[29] In Franz Kafka's novel *The Trial,* the protagonist Josef K. is arrested by an authority that never reveals the crime he has committed. In this living horror, K. finds himself at the mercy of a bureaucracy that works in secret, a law that is invisible, and a court that is untouchable. The book is a scathing satire of government bureaucracy and the totalitarian methods that would later define the Nazi and Soviet regimes of the twentieth century. K. descends into a surreal and tortuous rabbit hole: a hellish *Alice in Wonderland* from which there is no waking up. To prove fiction is always one step ahead of life, Kafka's cynicism toward modernity was tragically confirmed years later when his three sisters were sent to concentration camps during World War II. The term Kafkaesque has since come to define those situations in which people are trapped inside complex and nightmarish bureaucratic systems.

Much like Kafka, the sociologist Max Weber shared a profound ambivalence, even anxiety, about modernity. Weber famously used the phrase "disenchantment of the world" to describe a planet that had—since the scientific revolutions of the fifteenth century—chased away its magic and gods and replaced them with new deities: efficiency, rationality, and calculability. He observed that, slowly but surely, most human tasks—from birth to death—were becoming impersonally organized by bureaucracies. Weber feared these rigid structures would generate a society deadened by widespread conformity. Although Weber defined a bureaucracy in the way we imagine today (as offices composed of people, files, and rules),[30] he also argued it was a *disciplinary* power: "Bureaucracy develops the more perfectly the more it is 'dehumanized,' the more completely it succeeds in eliminating from official business love, hatred, and all purely personal, irrational, and emotional elements which escape calculation."[31]

Weber's doomful prophecy of a life enclosed by calculation has become a characteristic feature of technological civilization. Numbers and statistics saturate daily life. Weber called the rationalization of social life a type of "iron cage" or "steel hard casing" that imprisoned people "as never before in history."[32] This bureaucratic arrangement is soldered to an economy "bound to the technical and economic conditions of mechanized, machine-based production."[33] In turn, Weber adds, "War in our time is a war of machines."[34] A technical reality was gripping the human species. But even Weber could have scarcely foreseen the toxic dehumanization that would define the most brutal passages of the modern age as bureaucracy collided with technology and humanity slowly lost grip on its murderous creations. The twentieth century witnessed the ascent of technologies that could annihilate millions with the press of a button. Today, a hyperrational form of authority governs technological civilization, one based on impersonal and "objective" technics. The spread of this form of bureaucracy makes it difficult—or simply unnecessary—to *think*.

The "rule by Nobody" names this abstract system of control: a "tyranny without a tyrant," a form of rule in which "there is nobody left with whom one can argue."[35] The term belongs to Arendt, who was highly critical of the bureaucratization of everyday life. She writes that bureaucracy is "the rule of an intricate system of bureaus in which no men, neither one nor the best, neither the few nor the many, can be held responsible, and which could be properly called the rule by Nobody."[36] As Weber first poses, "The individual bureaucrat cannot squirm out of the apparatus into which he has been harnessed" because "he is only a small cog in a ceaselessly moving mechanism."[37] As such, by its very nature, bureaucracy is a threat to democracy, since a sociotechnical, hyperrationalized momentum takes hold. Arendt sees in Adolf Eichmann, the Nazi official who helped organize the Holocaust, the most profound kind of "desk murderer," a bureaucrat given unchecked power. Chalmers Jonson writes that this is "an equally apt term for George W. Bush, Dick Cheney, and Donald Rumsfeld—for anyone, in fact, who orders remote-control killing of the modern sort . . . say, the unleashing of a Hellfire missile from a Predator unmanned aerial vehicle."[38]

With the development of the drones, robots, and sophisticated machinery that operate upon the planet today, the rule by Nobody is now (perhaps fundamentally) a rule by technics. Arendt's definition can thus be stretched. The rule by Nobody is not only a bureaucratic form of control—rationalized, dehumanizing, and antidemocratic—but also profoundly *automating*. Societies administered under the rule by Nobody are *doubly enclosed*: enclosed by unyielding forms of rationality and enclosed by the technologies that process living matter. Artificial machines govern much of our speaking, acting, and killing in technological civilization. Under this rule by Nobody, we have become, according to Arendt, "thoughtless creatures at the mercy of every gadget which is technically possible, no matter how murderous it is."[39] The pieces of the rule by Nobody are certainly in place: secretive courts that rubber stamp distant assassinations, surveillance apparatuses that enclose the planet's communications, and robotic warriors that deliver freedom to unknowing suspects—Kafkaesque killing.

Such rationalized death management lies at the heart of the U.S. manhunt. According to reports that first surfaced in 2012, the program of targeted killings follows an extremely bureaucratic process centralized in the White House. The blueprint for pursuing terrorists is a targeting list called the "disposition matrix." As Greg Miller describes it, "The matrix contains the names of terrorism suspects arrayed against an accounting of the resources being marshaled to track them down, including sealed indictments and clandestine operations. U.S. officials said the database is designed to go beyond existing kill lists, mapping plans for the 'disposition' of suspects beyond the reach of American drones."[40] In a Google video President Obama defends the practice of drone killings in Pakistan saying, "It's not a bunch of folks in a room somewhere just making decisions." As it turns out, Obama is only partly right, since bureaucrats in a room somewhere are central to the targeted killing program. According to Gregory McNeal, this process has four main stages: identification, vetting, validation, and then nomination.[41] An individual is identified by his status with an organized armed group and his effectiveness within a network. This is a form of effects-based targeting in which an individual's social network is mapped. All of this information is stored within an Electronic Targeting Folder (ETF).

These ETFs populate the disposition matrix, a database that contains the names of dangerous individuals and the resources assigned to kill or capture them. This bureaucratic tool harmonizes the kill lists across the CIA and the Pentagon and embeds targeted killings within the National Counterterrorism Center (NCTC). Users of the database are guided by the White House's playbook, a document that details the guidelines for the use of extrajudicial force across the globe.[42] The target is then vetted and validated by a range of intelligence agencies, which analyze the tactical and strategic gains and losses associated with elimination. Finally, in the nomination stage, a vote takes place during a meeting of the National Security Council deputies—on what is known colloquially as Terror Tuesdays.[43] Here, a range of lawyers and officials, together with the president, view PowerPoint presentations of each potential target. The entire procedure resembles a law-enforcement model for scrutinizing criminal behavior, only the jury acts also as judge and executioner.

Once an individual is signed off for elimination, the actual job of killing begins. The process starts by fixing the target to a geographical site. Their location determines the next step: calculating the likelihood of collateral damage. Both the CIA and the military reportedly follow Collateral Damage Methodology (CDM), which allows analysts to visualize the blast zone of a proposed strike. This process has been computerized with FAST–CD (Fast Assessment Strike Tool–Collateral Damage), a piece of software that models collateral damage estimates. Relatedly, the Population Density Reference Table estimates the density of a population within the targeted blast zone. If collateral damage is unavoidable, then final authorization goes to a predetermined authority for proportionality assessment. If the Casualty Estimate Worksheet number exceeds the value of the Non-combatant Casualty Cut-Off Value (NCV), then a step known as Sensitive Target Approval and Review (STAR) is triggered. In such a case, either the secretary of defense or the president must sanction the STAR target.

The entire system, then, is rationalized at every step. From ETFs that contain detailed imagery and blueprints of target locations to weapons data that are tested with physics-based computer modeling,[44] the bureaucratic, even scientific, nature of targeted killings augments individual thinking with a machinic certainty. But this rationality does not translate into accountability in any meaningful way. After all, who is accountable? The software? The analyst? The drone? As David Kennedy describes the situation today, "Violence and injury have lost their author and their judge as soldiers, humanitarians, and statesman have come to assess the legitimacy of violence in a common legal and bureaucratic vernacular."[45] In a sense, accountability has been reduced to a quantitative exercise. Bureaucratic accountability under such conditions is self-referential, reminiscent of the enclosed cybernetic world of Vietnam's technowar.

"When we institutionalize certain things, including targeted killing," said a former deputy director of the CIA's counterterrorism center, "it does cross a threshold that makes it harder to cross back."[46] Harold Koh, former legal adviser to the Department of State, confided to a friend that stopping a targeted killing "would be like pulling a lever to stop a massive freight train barreling down the tracks."[47] Here, the problem of accountability is magnified. Accountability is not only reformatted to a quantitative exercise but collectivized to such an extent that no single individual in the kill chain is directly responsible. The personality of the CIA analyst, the drone pilot, or even the president is largely irrelevant, since the system swallows individual autonomy. This anonymizing effect alienates bureaucrats from the killing fields, such that the decision to take another's life is a technical concern. What is the rule by Nobody? It is an "apparatus," or an "architecture," replies Foucault, "for creating and sustaining a power relation independent of the person who exercises it."[48]

Technological Civilization, Part I: Control Societies

Schools, army barracks, factories, hospitals, asylums, and prisons were types of enclosure that defined the disciplinary societies of eighteenth and nineteenth century Europe. In each of these environments, individuals were subject to forms of state power that produced docile masses. Children were disciplined in school by an institutional hierarchy no less completely than were inmates languishing in jail. According to Gilles Deleuze, these sites produced "the organization of *vast spaces of enclosure.* The individual never ceases passing from one closed environment to another."[49] In the prison Foucault locates the most oppressive form of enclosure. The ultimate expression of this was Jeremy's Bentham's Panopticon, an architectural blueprint for a prison in which inmates could be watched at all times from a central tower. The enclosure served as a walled and watched space of state control. Moreover, the prison was "not alone, but linked to a whole series of 'carceral' mechanisms which seem distinct enough—since they are intended to alleviate pain, to cure, to comfort—but which all tend, like the prison, to exercise a power of normalization."[50]

For Deleuze the movement away from a closed space to a more diffuse system of enclosure represented a transition away from disciplinary societies to what he calls "control societies." And this movement had a spatial expression. Typical of disciplinary societies were enclosures designed to be discrete, self-contained, and bounded, or what Deleuze calls "molds." These molds are still present, of course, but control societies spiraled outward from these molds. A static spatial logic of enclosure was transformed into something more pervasive. "Enclosures are *molds,* distinct castings, but controls are a *modulation,* like a self-deforming cast that will continuously change from one moment to the other, or like a sieve whose mesh will transmute from point to point."[51]

Enclosure, in other words, evolved from a topographic spatial organization based on discrete sites like asylums into a topological system that was modulating and based on sociotechnological mechanisms of control. Like the disciplinary mechanisms of prisons, the enclosures of control societies are inherently carceral, even if their atmospheres are more diffuse. Populations remain policed on the inside, even if that inside is now foamy and fragmented, emergent and nonlinear, ambient and ecological. Within this mesh of biopolitical power, the human is undulatory, in orbit, and is barely able to retain its coherence. The individual becomes more of a *dividual*: an inchoate, divided subject endlessly chopped up and splayed among digital codes, algorithms, passwords, commodities, networks, CCTV cameras, credit card reports, e-mails, texts, biometric databases, avatars, videos, and social media. The dividual is the human subject of technological civilization, endlessly fragmented by modern control technologies.

The control society rests upon one technology in particular: the computer. The computer infects populations with codes that "mark access to information, or reject it." Disciplinary enclosures have been overrun by control mechanisms that do not require physical containment to generate widespread conformity. As Deleuze concludes, "What counts is not the barrier but the computer that tracks each person's position—licit or illicit—and effects a universal modulation."[52] The universal modulation effected by the Predator Empire is the condition of our times. In addition to targeting land, sea, and space, our psychosocial spheres are now subject to brutal processes of normalization. That the world is a battlespace is just another way of formulating this logic of endless control, in which the enclosure is topological rather than topographical and which targets the life of coisolated dividuals rather than communal individuals. As Hardt and Negri argue, "The Empire's institutional structure is like a software program that carries a virus along with it, so that it is continually modulating and corrupting the institutional forms around it. The imperial society of control is tendentially everywhere the order of the day."[53]

Software is the nervous system behind the control society. Fundamentally, the enclosure of human coexistence is sustained by the autonomous running of software code in the background of daily life. From the use of keycards to access locked doors in the smart city to the GPS systems used for navigation, software coordinates the existential grammar for the activities, behaviors, and spaces of control. Speaking to this, Martin Dodge and Rob Kitchin put forward the idea of "code/space" to describe the *production of space* by software and its intimate imbrication in everyday life: "Coded objects, infrastructures, processes and assemblages, and the technicity they engender, transduce space—beckon new spatial formations and spatiality into existence."[54] Under such an understanding, the landscape is transformed into a robotic ecology of technical events continually transducing human–nonhuman relationships. The ubiquity of computing now means few spaces are left not in some way impacted by digital technologies. "Most people in Western nations," write Dodge and Kitchin, "are living in a machine-readable and coded world—that is, a world where information is routinely collected, processed, and acted on by software without human intervention."[55] Software is driving automatic, automated, and autonomous enclosures, unburdening the human subject while empowering the rule by Nobody.

Indeed, the control societies that constitute technological civilization feed off a widespread, listless passivity: action, thought, and freedom are entirely unnecessary to the present order of things. A generalized political apathy numbs those cocooned in the enclosures of technological civilization. Politics of the fiercely agonistic, existential variety has been progressively enclosed by the rigid conformity or unblinking populism of control societies. This subjects human beings to economic models that reduce individuals to well-behaved animals whose sole purpose is to produce, consume, and die peacefully. Banal survivalism is the motto of such limp populations. For Alain Badiou these "atonic worlds" are what Western capitalism dreams of: places in which *nothing* happens. All that's left is "unreserved consumption and easy-listening euthanasia."[56] Atonic worlds are frictionless environments that human subjects float through in an unthinking slumber. Imagine a great shopping mall from which there is no exit: a sparkling interior in which one shuffles from CCTV camera to CCTV camera, unperturbed by the infrastructures that imprison body and mind.

Technological Civilization, Part II: Synthetic Shells

Targeted killing is merely the end point of a much wider technocratic system. The rule by Nobody administers the atonic worlds of control societies, endlessly modulating and dividing the human subject. It is therefore crucial to explore, in more detail, what exactly constitutes the human condition—and how the contemporary existential link between humans and machines is politically dangerous. A key guide to navigate this complex terrain is Arendt. She lists three activities fundamental to the human condition: action, labor, and work. Action, the rarest of all human activity, is defined as those political acts capable of disrupting the status quo. Laboring, on the other end of the scale, is a much more common part of the human condition and sustains the biological life of the species by producing consumable goods. Finally, Arendt defines work as the activity that creates the public worlds we inhabit.

Work, as opposed to labor, builds the artificial homes for humanity to live inside, or what Arendt calls "the artifice." The artifice is composed of those great civilizatory spheres that cocoon human existence, immunizing the species from the frost of the outside, blurring the lines between what is alive and what is artificial. Human existence is guaranteed by the stability and permanence of this nonhuman landscape that surrounds us. "The man-made world of things, the human artifice erected by *homo faber,* becomes a home for mortal men, whose stability will endure and outlast the ever-changing movement of their lives and actions, only insomuch as it transcends both the sheer functionalism of things produced for consumption and the sheer utility of objects produced for use."[57]

The artifice is not, however, a permanent feature of humanity, and neither is the work that supports it. Both of these existential-support systems are endlessly under attack by capitalism, which melts all that is solid into air and consistently devalues that which is not directly or indirectly profitable. The artifice is increasingly precarious—now more than ever. The Industrial Revolution began to replace the world-making craft of humans with mindless laboring. Consumption of these commodities, in turn, came to define us as a consumer society. As Arendt reasons, "We have almost succeeded in leveling all human activities to the common denominator of securing the necessities of life and providing for their abundance."[58] A society that rewards only labor is doomed to worldlessness, as the artifice rusts away and falls prey to mindless consumption and oppressive state control.

Of course, humans have always used artificial tools for their survival. Human evolution, in its broadest possible sense, is influenced by artificial systems as much as it is by DNA. As such, the forces that modulate the chaotic lives of individuals are not always natural. Contemporary ecosystems are infiltrated by immense technological systems: DNA can no longer be understood as the only evolutionary force. As the field of epigenetics has demonstrated, the environment is crucial in the development of the human genome, activating and deactivating genetic dispositions. A pure biology, understood in this way, is impossible. There is not only a flattening of the distinction between political and biological life in control societies but also a collapse between the organic and technology: biopower is always already a type of technopower.

Gilbert Simondon argues individuals are individuated through their interaction with artificial forces. Individuation fuses various social, psychic, and technical forces together. The last now conditions human subjectivity to an unparalleled extent. In the enclosures of technological civilization, we are surrounded by technical forces. Machines, according to Arendt, have "become as inalienable a condition of our existence as tools and implements were in previous ages."[59] For Stiegler this artifice, or what he terms "prosthesis," thereby "constitutes the reality of the human's evolution, as if, with it, the history of life were to continue by means other than life: this is the paradox of a living being characterized in its form of life by the nonliving."[60] As he continues, "It is in this sense that the *what* invents the *who* just as much as it is invented by it."[61] Technics is not reducible to technology as we imagine it today but is the broader prosthetic system of tools, languages, and materials that give human life its continuity through exteriorization. We inherit a past, a *what,* just as much as we inherit biological material. In this sense we also inherit an epigenetic infrastructure. The spiritual malaise we face stems from the impoverished relationship between the who and the what, between humans and technology.

The originary site of machinic alienation, the English industrial factory, gradually became a general condition of life inside technological civilization. "Machines demand that the laborer serve them," writes Arendt, "that he adjust the natural rhythm of his body to their mechanical movement."[62] Machines are no longer the means to realize humanity's highest ambitions. Rather, humans have become the means of realizing the machine's thoughtless, endless production cycles. This unchecked production for production's sake reaches its limit with the complete automation of activity. The existential danger, however, is neither simply the artificialization of the human spirit nor the intensification of mindless consumption but the erosion of the durability of the world, of the artifice. As Arendt insists, "The question therefore is not so much whether we are the masters or the slaves of our machines, but whether machines still serve the world and its things, or if, on the contrary, they and the automatic motion of their processes have begun to rule and even destroy world and things."[63]

With the sustained enclosure of the public artifice by globalized commodity circuits, a profound change in the human condition is taking place. The public sphere is a vital constituent of reality, according to Arendt: it is where we are seen and heard, where our sensible experience as living, breathing humans is affirmed. It is, in other words, where *existence as coexistence* is validated—a precarious and primordial loop. If we lose the public sphere, the commons—and make no mistake, we *are* losing it to a fissiparous twenty-first-century enclosure—then we lose the world itself. Without a public artifice, a shared world to dwell inside, our participation in the immortal activity of building a common home on the planet is destined to be supplanted by fragmented enclaves of private capital. Technological civilization becomes the universal space of dissociation: a matrix of atonic worlds that float in the nether.

Of course, technological civilization provides compensatory practices for endemic worldlessness. We have fled to our private spheres to ever greater extents, erecting bubbles that immunize us against our own nihilistic tendencies. Indeed, as mass society is further and further disenchanted, these private orbs become ever more enchanted. But commodities do not provide a world, much less anchor one. An endemic loneliness washes over consumers sealed inside their comfort capsules. Furthermore, this retreat into the private sphere harbors a *totalitarian* tendency. The enclosure of the public sphere replaces a plurality of perspectives with a singular, hegemonic way of thinking and acting. As humanity flees into its private shells, the public artifice is hollowed and dominated by a logic born of the marriage between capitalism and state power. "It is quite conceivable that the modern age—which began with such an unprecedented and promising outburst of human activity—may end in the deadliest, most sterile passivity history has ever known."[64] Perhaps, the only action left for modern dividuals is to relax as the civilizatory conveyor belt reliably tugs us toward a comfortable grave.

The existential emergency we face is real and pressing: at its most dangerous level, humans dissolve entirely into the machines that enclose them. What would this look like? If an observer were to enter Earth's upper atmosphere and peer down at the globe below, what kind of human activities would they discover on the surface of our planet? As Arendt answers, perhaps our busy—but empty—lives "would appear not as activities of any kind but as processes, so that, as a scientist recently put it, modern motorization would appear like a process of biological mutation in which human bodies gradually begin to be covered by shells of steel. For the watcher from the universe, this mutation would be no more or less mysterious than the mutation which now goes on before our eyes in those small living organisms which we fought with antibiotics and which mysteriously have developed new strains to resist us."[65]

When the artifice no longer shelters its denizens and becomes the hollowed space of commodity circulation, society becomes terminally disaffected. Technological civilization is so unbearable for so many because it rests upon this endemic wordlessness.

Technological Civilization, Part III: Disaffected Individuals

As technological civilization expanded, the humans it housed became increasingly alienated. The first stage of this world alienation was the enclosure of the commons, which expropriated people from their own means of survival. Second came the replacement of a durable artifice, a public world, with a civilization obsessed with instrumentalizing everything, setting all that was solid into motion. Finally, the earth itself became progressively alienated as an object of science. Ever since Galileo peered into his telescope, the earth has been grasped from the perspective of a universal science. It is now grasped from the perspective of a *universal war.*

The Predator Empire, which privileges full spectrum dominance, is constantly remaking the earth into a unified battlespace with discrete targets, one policed as a single control society. As Arendt argues, "Just as the family and its property were replaced by class membership and national territory, so mankind now begins to replace nationally bound societies, and the earth replaces the limited state territory."[66] While the planet may be profoundly connected by modern communications, this connection comes at the cost of "alienating man from his immediate earthly surroundings."[67] The contradictions of mass expulsion are endlessly felt.

In 2013, the year the video game *Call of Duty: Ghosts* made over $1 billion in twenty-four hours, around forty thousand Americans took their own lives. While millions were plugging into technical civilization, basking in the warm glow of its screens, thousands of human beings were permanently unplugging from a civilization that no longer sheltered them. At the same time, the wealthiest country on earth had 2.4 million people incarcerated in 1,719 state prisons, 102 federal prisons, 2,259 juvenile correctional facilities, and 3,283 local jails.[68] Compounding these depressing statistics, children with an incarcerated parent were more than three times more likely to suffer from depression and other behavioral issues.[69] The economic and psychological misery of surplus populations, of course, provides a new site of enclosure for the prison–industrial complex, which profits directly from a society of disaffected individuals.

A great deal of our human misery swells from the automated lives we now lead. The technical world is always one step ahead of society, which must constantly adapt to the psychic shocks brought about by automated ways of being-in-the-world. Technology, as Stiegler notes, is productive of a social *disequilibrium*. Within the next two decades, economists predict nearly half of today's jobs in the United States will be automated.[70] With each worker replaced by machine, the economic contradiction between automation and social well-being is exacerbated. These surplus populations have less and less invested in a technological civilization that no longer benefits them. Indeed, civilization incessantly breaks the pact it made to guarantee its denizens a better life, and with that pact broken, trust in the system evaporates.[71] The development of a technological system that operates against society is what Stiegler calls "irrational." This contradiction, of course, is a major source of instability. But even as the technical system becomes more irrational, it develops capacities to quell unrest. As crises continued to develop, "it would inevitably lead us to a *state system of totalitarian terror, a politics of terror* to counter the terror of those in despair."[72]

The collective phenomenology of the control society is formed by mass digital modulations that probe deep beneath the psychic skin of the individual and scramble its drives, creating armies of electronically lobotomized dividuals. Sigmund Freud observes an individual's superego is first formed through an emotional identification with his or her parents: the child unconsciously adopts the characteristics of its most intimate authority. Stiegler argues today's hypercapitalism *liquidates* this primary identification. Mass media and marketing target children from a very young age, saturating their attention with commodities, "thus provoking indifference towards their parents and to everything around them."[73] But that's not all. The secondary identifications of adulthood are also under attack: capitalism targets those social structures—the artifice—that provide a sense of worldly stability and attachment. The dividual, already fragmented by cerebral marketization, finds itself floating in an atonic world of impermanence, awash in an ocean of social flotsam and jetsam. The civilizatory artifice is fractured by the unyielding circulation of capital.

Indeed, by enclosing the world with twinkling commodities and drowning consciousnesses in digital simulacra, capitalism directly exploits the process of individuation. Psychotechnologies short-circuit and divert desire toward consumption. One of the early founders of this form of existential marketing was Edward Bernays, Freud's nephew and an important figure in public relations. As Bernays argues in his 1928 book *Propaganda,* "As civilization has become more complex, and as the need for invisible government has been increasingly demonstrated, the technical means have been invented and developed by which opinion may be regimented. With the printing press and the newspaper, the railroad, the telephone, telegraph, radio and airplanes, ideas can be spread rapidly and even instantaneously over the whole of America."[74] This invisible government is the background matter of our social lives, the pervasive rule by Nobody that modulates various affective states. It is a type of infrastructural psychopower.

If the artifice offers a space of individuation, a world capable of nurturing primal narcissism (and affirming existence as coexistence), then enclosure shuts it down, replacing the commons with the enclave and the individual with the dividual. Without a supportive world to live in, a place capable of nurturing individuation, exiled dividuals fall prey to the unworldliest of drives, seduced by regressive forms of extremism. According to Stiegler, absolute disenchantment is the erosion of all solidarities necessary for society and the ascent of purely instinctive drives—the Freudian death drive, in particular.

In other words, if technological civilization fails to support or nurture human individuality, then it *disindividuates*. The more the individual of the control society is unburdened of its ability *to be an individual* (through thinking, acting, and communicating with others in a common world), the greater the *loss of self*. In this sense the modern age is defined by a loss of coexistence, "an immense process of psychic and collective disindividuation."[75] Individuals lose their own existences—they lose themselves *as* selves through a *technical disinheritance*. Often denounced as individualistic, the modern age loses selves at a staggering rate, producing herdish dividuals everywhere. Dividuals are thus doubly exiled: exiled from a common world and exiled from their own sense of individuality.

Hypercapitalism thus creates the conditions for what Stiegler calls a generalized "proletarianization" of society.[76] This names the stripping of the human of its worldly skill (or what Arendt would call "work") and the installing of an epigenetic system that automates human behavior. Since the enclosure of the commons, producers have long had their way of life massified. Today, consumers are further robbed of their singular way of being-in-the-world. Psyches are synchronized, and behavior is computationally modeled, predicted, and controlled. As Stiegler argues, "In this way, capitalism, in its hyper-industrial—that is, hyper-computational—stage, expresses a *totalitarian tendency* consisting in the tendency to reduce everything to calculation, to turn *all singularity* into mere *parts of a whole*."[77] Human singularities are everywhere turned into points identified by their location within a database: a hospital record, a credit card record, a disposition matrix, or whatever is their identifying code. Humanity's adoption of technics is not simply a one-way relation. Technics soon come to domesticate humanity: the what assails the who, and the dead seeps into the living as its very condition of being.

Biopower, taken to its limit, is not simply the control of the body, the population, or even the human species but the control of the technical ecosystem by which individuals *become*. The Leviathan absorbs the psychological resources of humanity. Every woman made homeless by a predatory capitalism, every man incarcerated by a broken war on drugs, and every patch of the planet exploited by a splintering global elite strangles authentic sources of power by eroding the human artifice of its underlying vitality. If individuation is the process by which the mental interior binds to the technical exterior, that exterior—what Stiegler calls a "retention system"—is the locus of a profound (geo)political struggle. In this sense, the formation of the human psyche is a crucial site of control. Biopolitics thereby soon becomes biotechnics as the population is phenomenologically enslaved by forms of state-sanctioned psychopower. "Now," writes Stiegler, "the liquidation of the super-ego causes societies to become police-societies."[78] Today, in the age of the control society, biotechnics is principally computational and seeks to "eliminate those singularities that resist the calculability of all values on the market of economic exchange."[79]

If technological civilization fails to control its own contradictions—namely, mass expulsion—its very existence is threatened. The Predator Empire must therefore dominate the coordinates—real and virtual—of human individuation. The extremely unequal control societies we live in today, which in the past may have generated rebellion, are policed by sophisticated technical systems and swarms of psychic apparatuses. The more dissociated and disinvested the form of authority, the more it comes to rely on violent apparatuses of control. Indeed, the so-called state apparatus is today being winnowed down to its most basic, violent function: to enforce security with nonhuman mediators.

As Arendt warns, those governments who feel power slipping from their hands "have always found it difficult to resist the temptation to substitute violence for it."[80] And this triumph of violence over the power of the people is "never possible without instruments."[81] These instruments of violence have changed throughout history, from police batons to helicopters. "Only the development of robot soldiers," writes Arendt, "would eliminate the human factor completely and, conceivably, permit one man with a push button to destroy whomever he pleased."[82] The grim irony here is that decades after Arendt penned this sentence, the world is now infested with a range of "robot soldiers." Their presence—on land, on sea, and in space—threatens to tip the balance of violence over authentic sources of power, of the machine over the human, of totalitarianism over democracy.

Sleepwalking into Totalitarianism

The twentieth century was a century of totalitarianisms. From the punishing Soviet Gulag network to the vicious Nazi concentration camps, enclosure, exile, ghettoization, and incarceration were blueprints for social control across much of Europe. As Anne Applebaum writes, "In Stalin's Soviet Union, the difference between life inside and life outside the barbed wire was not fundamental, but rather a question of degree."[83] The passage into the twenty-first century has renewed and transformed the space and logic of enclosure. The threat posed by totalitarianism, however, has remained alive and well. The mass technical disindividuation of modernity continues to shatter old social bonds, and the apparatuses that swarm across the globe to police the resulting psychic breakdowns are inherently totalizing.

What unites millions of us in technological civilization today is mass alienation, a condition fed by the ongoing enclosure of the planet. Alienation is a kind of homelessness: the very fabric of the world, the guarantor of *existence as coexistence,* is ripped apart. Totalitarianism, in turn, parasites on the worldless dividuals that go to work every day like lost phantoms stuck on repeat (and those are just the fortunate ones). Arendt calls this "mass" a "terrifying negative solidarity" of isolated people. The "mass man," as she explains, describes the lonely and atomized dividual that has given up caring about politics. A timid creature, the mass man is concerned only with safeguarding his private sphere, having long abandoned the commons and politics. The eruption of totalitarian regimes in twentieth-century Europe was enabled by the existence of this mass, according to Arendt, and was organized most effectively by Adolf Hitler in Nazi Germany. "Totalitarian movements are mass organizations of atomized, isolated individuals."[84] Totalitarianism offered the exiled masses a place to call home: it invited them to reintegrate into a space of meaning where they could escape from modernity's cool nihilism and rootlessness. A fictitious reality, no matter how artificial, is better than the feeling of hollow superfluity. For Arendt, Nazi propaganda was so successful because it restored mass man's self-respect and worldly stability. This was a form of total domination, understood as the organization of dividuals into a kind of unified pan-individual.

At the foundation of the Nazi machine stood the concentration camp. "The camps are meant not only to exterminate people and degrade human beings, but also to serve the ghastly experiment of eliminating, under scientifically controlled conditions, spontaneity itself as an expression of human behavior and of transforming the human personality into a mere thing, into something that even animals are not."[85] After heads were shaved, uniforms were worn, and families were ripped apart, the human being was alone, utterly alone. "Here, murder is as impersonal as the squashing of a gnat."[86] The Nazi regime converted human *action*—which for Arendt is the task of starting "something anew"—into "bundles of reactions that behave in exactly the same way."[87] This is the citizen of the totalitarian state. Totalitarianism destroys the fabric of the world by enclosing the most intimate and psychological spaces of human existence, thereby imposing a form of total and inescapable terror.

Might the problem of good and evil, then, be connected with our ability to think—and act—freely and independently? Arendt attracted a storm of controversy after covering the 1962 trial of SS official Adolf Eichmann in Jerusalem. Eichmann was a major figure responsible for putting millions of Jews to death, organizing their mass deportation to ghettos and concentration camps. Arendt characterizes Eichmann not as a monster but as an unthinking bureaucrat. "The trouble with Eichmann," laments Arendt, "was precisely that so many were like him, and that the many were neither perverted nor sadistic, that they were, and still are, terribly and terrifyingly normal."[88] Like so many in the Third Reich, Eichmann collapsed his own actions, his ability to think, inside the reality engineered by the Nazi regime. This was a national hallucination that shielded so many from *thinking* about the barbarous acts that were committed. Executions by gas chamber supported the lie that the Final Solution was a medical project and not plain murder, just as the Action T4 program "euthanized" tens of thousands of mentally sick Germans.

At each step of the downward spiral into the Holocaust, reality was veiled by widespread, accepted ideological euphemism. "As Eichmann told it, the most potent factor in the soothing of his own conscience was the simple fact he could see no one, no one at all, who was actually against the Final Solution."[89] A whole nation was plunged into thoughtless slumber. Upon his death Eichmann rehearsed a set of clichés, as if his conscience was unperturbed by the trial. "It was as though in those last minutes he was summing up the lesson that this long course in human wickedness had taught us—the lesson of the fearsome, word-and-thought-defying *banality of evil.*"[90] Arendt's famous term highlights the interdependence of thoughtlessness and evil, which are often wrapped in those administrative massacres carried about by the state. Could the activity of thinking really "make men abstain from evil-doing or even actually 'condition' them against it?"[91]

Greek philosopher Socrates is credited with saying, "The unexamined life is not worth living." For him the activity of thinking provides the very oxygen of human being. It shakes the established knowledges by which humans orient themselves and rips them away from their dogmatic slumber. Unlike the accumulation of knowledge, which asks for the what or the how, thought poses an altogether more fundamental question, Why? Danger, therefore, is posed to the establishment whenever *thought thinks the unthought*. As Arendt puts it, "There are no dangerous thoughts; thinking itself is dangerous."[92] Thinking is a form of withdrawal from the immanent order of the world: it is "always out of order" and "interrupts all ordinary activities."[93] As Badiou adds, "All resistance is a rupture with what is. And every rupture begins, for those engaged in it, through a rupture with oneself."[94] To say people are capable of thinking is to say they are capable "of prescribing a possible that is irreducible to the repetition or the continuation of what exits."[95] Whenever we engage in thought, the reality around us is suspended, and the logic of the world is questioned. Thinking is thus a silent dialogue with oneself, what Arendt calls the "two-in-one," and it is the first site in which human plurality is established: between me and myself. Thinking, in its most elemental sense, is the adoption of the viewpoint of another, and without it we cannot possibly hope to judge between right and wrong.[96]

A human collective without thought is doomed to follow a pathway to its own destruction. And for the most part technological civilization doesn't *require* us to think. By and large, it doesn't even *want* us to think. Thinking is superfluous to the fast circulation of commodities and the instant acts of consumption: it is wasted energy. The two-in-one of our inner dialogue is replaced by the immersive, dazzling one of technological civilization—an immanent relationship that removes the space (materially and psychologically) for dialogue or critique.

Western democracy is founded on and gains political authority from its own endless self-critique. Without this process democracy soon flips into totalitarianism. As Arendt writes, "By shielding people from the dangers of examination, it teaches them to hold fast to whatever the prescribed rules of conduct may be at a given time in a given society."[97] Inside every totalitarian regime lies this impulse to replace thought with obedience. Ideological thinking is precisely this form of total explanation that provides a logical consistency to the world. The insecurity of thinking is replaced with the security of logic, and the insecurity of living is replaced with the security of surviving. The creative and unruly spaces of plurality are under profound physical and psychological enclosure by the Predator Empire.

Cyberconflict

Multiple spaces of human coexistence have been enclosed under the Predator Empire, and as argued, the Internet connects many of these together. The recent history of cyberconflict exemplifies the insecurities and contradictions born of a global synchronization of electronic activity. Broadly speaking, cyberspace is the online environment produced by the Internet's underlying hardware. Additionally, cyberspace covers the systems that regulate critical civilian infrastructure. This includes millions of supervisory control and data acquisition (SCADA) systems that regulate, for example, the gas and water supply entering a city. In the future these will be sites of increased risk. Between 2011 and 2013, the number of times critical infrastructure of the United States was probed or intruded upon increased 1,700 percent.[98] The U.S. military is now immersed in a continuous cyberconflict: one without spatial or temporal borders and one that blurs the lines between politics, peace, and war. "In cyberspace," write Peter Singer and Allan Friedman, "an attack can literally move at the speed of light, unlimited by geography and the political boundaries. Being delinked from physics also means it can be in multiple places at the same time, meaning the same attack can hit multiple targets at once."[99]

Cyberspace has buttressed and eroded state power simultaneously. On the one hand, cyber capabilities reinforce traditional military power and extend the spaces of surveillance. On the other hand, a handful of tech-savvy individuals now possess enormous geopolitical influence. In 2013 the company McAfee reported new malware was being released every second. This malware includes botnets: software programs that string together millions of compromised computers to form "zombie hordes" capable of attacking a network. Advanced Persistent Threats (APTs), conversely, involve highly specialized cyber teams infiltrating and exfiltrating a system for personal data. In 2009 researchers discovered an APT called GhostNet, a surveillance infrastructure that had infiltrated computers in over one hundred countries, including systems belonging to foreign governments. As Richard Clarke and Robert Knake explain, the twenty-two-month operation "had the ability to remotely turn on a computer's camera and microphone without alerting the user and to export the images and sound silently back to servers in China."[100] Indeed, China is often singled out as the main source of global cyber attacks, but this masks a complicated range of actors: hacktivist groups regularly jostle with quasi-state cyber militia.

U.S. intelligence specialists conducted 231 cyber operations in 2011—a number that has surely grown since then.[101] Leon Panetta, the former CIA chief who helped launch Stuxnet, warned the United States faced a future "cyber Pearl Harbor." In October 2012 Obama signed Presidential Policy Directive 20, which cemented the role of distinctly *offensive* cyber operations. The nerve center of the U.S. military's cyber operations is CYBERCOM, a unified command formed in 2010. As with land, sea, air, and space, cyberspace is treated as an operational domain. General Alexander, speaking ahead of CYBERCOM's creation, argued the United States needed a force that could proactively retaliate against foreign attacks. "The only problem is that the Internet—by its very nature—has no borders and if the United States takes on the mantle of the world's police; that might not go down so well."[102] In 2013 Alexander revealed thirteen teams of computer experts had been established to strike back against cyber attacks. So far, the use of cyberspace for effecting physical terrorist attacks has been nonexistent, although it is regularly used for disseminating extremist propaganda.

Unsurprisingly, governments are anxious about cyberspace. As one White House report warns, "Consequently, a growing array of state and non-state actors are compromising, stealing, changing, or destroying information and could cause critical disruptions to U.S. systems."[103] Businesses are worried, too. Ninety-seven percent of Fortune 500 companies have been hacked.[104] Thus, the idea of a new worldwide cyberconflict is based on the growing digitization—and policing—of criminal and civilian activity. As Singer and Friedman lament, "A uniquely democratic space created for communication and sharing is instead being transformed into a future battleground."[105]

China, for example, has built a national firewall (the Golden Shield Project) to protect its Internet sovereignty, and Russia has proposed a similar Internet within an Internet. Both digital enclosures allow the government to censor the exchange of online opinion. Indeed, in 2010 a Pentagon official called for the United States and Europe to erect a NATO cyber shield modeled after Cold War–era nuclear missile defenses. The idea of cyberconflict is not simply a clash between nations, then, but a fight to secure (or synchronize) the minds of populations and immunize them against potentially subversive thoughts. In one report on Internet freedom, thirty-four of the sixty countries assessed experienced a negative trajectory toward widespread attempts to censor freedom of speech.[106]

Cyberwar is the most serious form of cyberconflict, in which the aim is to cause physical damage and disruption. An early example of this was in 2007 when Estonia's entire Internet infrastructure came under attack from cyber militia. These attacks—part of what some call Web War One—were reportedly sanctioned by Russian secret services.[107] A similar pattern was repeated during Russia's conflict with Georgia in 2008, when multiple botnets attacked Georgian systems.[108] China has similarly invested heavily in cyberwar capabilities. The People's Liberation Army's Unit 61398—the Comment Crew, or Shanghai Group—is reportedly involved in gathering political, economic, and military intelligence on U.S. organizations. In 2013 it was caught allegedly hacking into multiple American corporations and infrastructural systems.[109] Cyberwar can also be folded into more conventional war, as when Israeli hackers compromised Syria's air defenses in 2007 or when the U.S. military hacked Iraqi IT infrastructure to send warning e-mails to military officers.[110]

In 2008 the United States began one of the most well-known cyber attacks. A band of cyber warriors in the NSA developed a software program—later known as Stuxnet—that targeted Iran's nuclear centrifuges. The site was Iran's secretive Nantaz laboratory, which existed in a secure air gap because it was not connected directly to the Internet. This meant any attempt at disrupting the centrifuges had to enter the system through a removable USB hard drive. Subsequently, an electronic worm was developed that could jump from a hard drive and tunnel into the mainframe. Initially, it was a success. Iranian scientists looked on puzzled as centrifuges shattered around them. But trouble hit in the summer of 2010 as the virus escaped onto the Internet and infected computers around the world. While Stuxnet was not a crippling setback for Iran, it set a geopolitical precedent. It was the first virtual weapon to cause physical damage. Many fuzzy geographical boundaries spiral out from this situation, including whether a cyber attack is an act of war.

We are standing at the dawn of cyberconflict. But it's a conflict that will range from data theft to physical damage. As John Arquilla and David Ronfeldt wrote in a seminal 1993 article, "Cyberwar takes a different view of what constitutes the 'battlefield.' Cyberwar depends less on the geographic terrain than on the nature of the electronic 'cyberspace.'"[111] As the Internet of Things populates this expanding cyberspace, the risks of domestic disruption and other, more intimate forms of violence only increase.

Another unexpected battlefield is cognitive space. As discussed, cyberspace is an important site of individuation. One UK military project under development, Full Spectrum Targeting, "measures future battlefields in social and cognitive terms rather than just physical spaces. Emphasis is put on identifying and co-opting influential individuals, controlling channels of information and destroying targets based on morale rather than military necessity."[112] Likewise, for years the U.S. military has been developing tools to accurately record, analyze, and anticipate mass civil breakdown. Following the global banking crisis in 2008, the U.S. Department of Defense began the Minerva Research Initiative to understand the social, cultural, and political forces responsible for international civil unrest. One such Minerva project investigates the role of social media in protests.[113] In short, cyberconflict collapses biological life and technical power, uniting neurons and electrons. In this sense, cyberspace is fast becoming yet another domain to be policed by the Predator Empire.

The NSA, Part I: SHAMROCK

The NSA is the premier military signals intelligence agency for intercepting, collecting, and decrypting foreign (and domestic) communications. It was established by Harry Truman in 1952, and its purpose was to consolidate the cryptologic units from World War II and provide the executive branch with foreign intelligence. Employing tens of thousands of analysts today, the NSA is the largest intelligence branch in the world and is headed by a director that also sits at the top of CYBERCOM. It is headquartered at a 1.8-million-square-foot complex at Forde Meade that cost $3.2 billion and is fed by its own 150-megawatt power station.[114] The NSA's surveillance infrastructures include geostationary satellites, listening posts that tap into telecommunications, satellite downlink stations, data storage facilities, and of course the headquarters in Maryland, where supercomputers perform automated cryptanalysis. The ubiquity of cyberspace in our daily lives has vastly increased the spaces and possibilities of surveillance. With technology replacing the role of spies and informers, the state's watch over its population is made far easier. No government organization, let alone surveillance agency, has ever held this much surveillance power in human history. Fed by an annual black budget of at least $10 billion, the NSA intercepts the communications of more than a billion people worldwide, tracks the geolocation of millions of cell phone users, and captures data about American citizens within its electronic dragnet. It's not just about national security either. The NSA has been reportedly involved in economic espionage, diplomatic spying, and audio surveillance waged against entire countries.

The very existence of a surveillance apparatus, big or small, poses an inherent danger to freedom. The threat of being watched—anywhere, anytime—changes human behavior and, worse, our ability to think and express ourselves freely. The Stasi (Ministry for State Security) in East Germany was one of history's most dreaded examples of a repressive intelligence and secret police network. The notion that U.S. surveillance could somehow resist the same downward spiral that dragged Europe into a living nightmare is counter to even recent American history.

As covered in chapter 3, in the 1970s Senator Frank Church documented a history of widespread spying on U.S. citizens, beginning with so-called subversives in the Cold War. FBI director J. Edgar Hoover ran a domestic counterintelligence program known as COINTELPRO, which began in August of 1956 and opened files on more than a million Americans. Anybody and anything that threatened the prevailing order—the civil rights movement, environmental protestors, antiwar activists, even Martin Luther King Jr.—was targeted. As the Church Committee stated, beginning in 1956 "the Bureau conducted a sophisticated vigilante operation aimed squarely at preventing the exercise of First Amendment rights of speech and association, on the theory that preventing the growth of dangerous groups and the propagation of dangerous ideas would protect the national security and deter violence."[115] But the FBI wasn't alone in targeting domestic enemies. In a parallel program the CIA created over seven thousand files on U.S. citizens under Operation CHAOS, which targeted the student antiwar movement during the Vietnam War.

The Church Committee discovered the NSA had intercepted millions of private telegrams sent and received by U.S. persons to foreign destinations, dwarfing even the CIA's mail-opening program. The origins of the SHAMROCK operation lay in the wartime censorship laws of World War II, which allowed the Pentagon to review international message traffic. Between August 1945 and May 1975, Project SHAMROCK was "probably the largest government interception program affecting Americans ever undertaken," according to the Church report.[116] No warrants were needed, and up until 1975, the NSA had no congressional oversight. Few in the NSA even knew of its existence, despite the agency intercepting 150,000 telegrams each month with the cooperation of three international telegraph companies. By the time magnetic tapes were introduced in the mid-1960s, "telegrams of citizens whose names were on NSA's 'watch list' could be selected for processing by NSA analysts."[117] This was, in other words, the dawn of automated surveillance.

Between 1967 and 1973, information obtained via SHAMROCK on U.S. citizens was used to form a watchlist known as MINARET. This list was passed on to the CIA, the Secret Service, the Bureau of Narcotics and Dangerous Drugs, and the Department of Defense. Hoover wrote of MINARET, "Information derived from this coverage has been helpful in determining the extent of international cooperation among New Leftists."[118] For the most part, the threat to national security was understood as civil rather than criminal, which meant the MINARET watchlist was a dangerous form of political policing. Nearly all of the elements of contemporary NSA spying can be located in the precedents set by SHAMROCK.

The revelations of domestic espionage uncovered by the Church Committee led to the passage of the Foreign Intelligence Surveillance Act of 1978. This brought surveillance under the specially created Foreign Intelligence Surveillance Court (FISA Court). Located within the Department of Justice, it acts as the main hub for issuing surveillance warrants. Warrants are required, however, only if one party of surveillance is American. If both parties are foreign, then NSA surveillance falls under Executive Order 12333, meaning the NSA is free to watch anybody. For the most part, the FISA Court has been a compliant court. For its first twenty-four years, between 1978 and 2002, it rejected zero government requests. In the remaining decade it rejected only eleven. As Greenwald concludes, "From its inception, FISA has been the ultimate rubber stamp."[119]

Even this form of judicial oversight came under attack after Congress passed the USA PATRIOT Act in October 2001. Under its Title II authorities, or Enhanced Surveillance Procedures, FISA was modified in two important regards. First, section 206 permitted roving wiretaps able to hop across multiple telecommunication lines. Second, section 215 enlarged the scope of materials that could be collected to "any tangible things."[120] This was initially known as the "library provision," since it granted the government wholesale permission to target any "business" record belonging to a person—library card usage, tax records, Internet usage, credit card information, medical records, and so on.

Furthermore, section 215 lowered the legal threshold for a court warrant to be issued. Instead of needing specific and articulable facts that a target posed a direct risk, those individuals who were merely relevant to an investigation could be targeted. As the FISA Court admitted in a redacted 2013 ruling, "Accordingly, now the government need not provide specific and articulable facts, demonstrate any connection to a particular suspect, nor show materiality when requesting business records under Section 215."[121] The burden of proof is simply an undefined relevance rather than probable cause. Here, the relevancy clause is interpreted broadly. The logic behind section 215 is that everybody's records are relevant because anybody could be a terrorist. Section 215 would later be used by the U.S. government as justification for the NSA's widespread surveillance of Internet traffic.

President Bush went beyond PATRIOT Act provisions when he signed a Presidential Directive in 2002 that allowed the NSA to "eavesdrop on Americans and others inside the United States to search for evidence of terrorist activity without the court-approved warrants ordinarily required for domestic spying."[122] Crucially, this special collection program took place *outside* the FISA Court process and was thus in contravention of congressional legislation that stated FISA was the exclusive means for this type of electronic surveillance. By the time this warrantless eavesdropping story broke in 2005, around five hundred Americans and seven thousand foreign suspects were being watched. As it turns out, Bush's NSA metadata collection was part of a bigger operation known as Stellar Wind, which reportedly continued until 2011.

Controversy over the Bush-era warrantless wiretapping did, however, lead to the passage of more congressional legislation. But all the 2008 FISA Amendment Act did was effectively legalize the extrajudicial espionage that preceded it. The NSA must still obtain a court-approved warrant for targeting a U.S. person, but everybody else in the world is fair game (including foreign heads of state), even if they are communicating with U.S. persons. Under section 702 of the 2008 law, the NSA is required merely to submit to the FISA Court its general guidelines before receiving blanket authorization for eavesdropping on the planet's population. In June 2015 the Senate passed the USA Freedom Act, ending the bulk interception of U.S. telephone records and requiring phone companies rather than the NSA to store metadata. Many praised it as the most significant reform since 1978, while detractors argued it didn't go far enough. Both were correct.

The NSA, Part II: The Snowden Revelations

Much of what we know about the NSA stems from the documents leaked by former contractor Edward Snowden in June 2013. The surveillance systems employed by the NSA could have therefore changed significantly in the past years. In any case, the documents indicate by mid-2012 the agency was already processing twenty billion Internet and telephone records from around the globe each day.[123] Indeed, hundreds of millions of devices are part of the agency's database. Much of this processing is now done automatically with sophisticated computer software like SHELLTRUMPET, a near real-time analyzer that produces tips for other NSA units.

The NSA has used multiple methods to intercept communications traffic. First, the NSA can request metadata records from telecommunication companies, as it did during Project SHAMROCK. The agency can also request information from foreign spy agencies like GCHQ (Britain's premier spy agency, the Government Communication Headquarters), which allows the U.S. government to bypass domestic legal restrictions (and vice versa). If that fails, the agency can tap directly into fiber-optic telephone lines. This wiretapping is known by the agency as upstream surveillance. Since the 1994

Communications Assistance for Law Enforcement Act, U.S. telecom providers have been legally obligated to construct their networks with wiretapping capabilities.[124] There are at least ten major SIGADS (Signals Intelligence Activity Designators) where the NSA and its partners actively tap communication networks. As more information is routed through Asia–Pacific networks, the NSA is reportedly buying real estate across the planet to extend their hacking capabilities. Under a program called RAMPART-A, Australia, Canada, and New Zealand have all provided the NSA access to their country's infrastructure. By 2013 the NSA had constructed at least thirteen RAMPART-A sites that mined data from seventy different cables, allowing it to tap into "congestion points around the world."[125] Three terabits of phone calls, e-mails, and other online data are intercepted every second in this way.

There are also more creative methods for hacking the planet's conversations. NSA hackers have compromised foreign computers using sophisticated malware (called Quantum Insertion) to exfiltrate personal data. Up to one hundred thousand computers around the world have been infected this way.[126] Indeed, under a $652 million[127] campaign of clandestine activity codenamed GENIE, around 1,800 NSA hackers exploited foreign IT infrastructures belonging to Iran, Russia, China, and North Korea. The Office of Tailored Access Operations (TAO) is the powerful NSA group charged with developing these unique digital implants. This sophisticated software can tunnel into thousands of connected networks, lying dormant for months to create back door entrances for future access. By the close of 2013, GENIE was projected to control at least eighty-five thousand implants across the world. The NSA's Access and Target Development department has also intercepted U.S. computer hardware before it was exported to other countries, implanting back door surveillance hacks.[128]

If hacking directly into computer hardware fails, then why not hack into corporate servers? PRISM was the name of a $20 million program that enabled the NSA's Special Source Operations division direct access to the servers of nine telecommunication giants, including Google, Facebook, Yahoo, and other U.S. technology firms.[129] Over seventy-seven thousand intelligence reports have cited the PRISM program as their source. Many U.S. companies initially cooperated with the NSA, and it was later reported some had accepted millions of dollars of government money for "compliance costs."[130] Given that hundreds of millions of people use these services, PRISM granted the NSA unfettered access to the private lives of an enormous population. Where the cooperation of telecommunication providers was not as forthcoming, the NSA developed ways to kick down the back door. Under the MUSCULAR program, jointly operated with GCHQ, the NSA hacked into Yahoo and Google data centers across the world, positioning itself at a crucial intercept point between the public Internet and corporate storage clouds.[131]

The NSA has employed sophisticated software programs to sift through all the data it collects. Since 2007, X-KEYSCORE has been the NSA's program to organize the e-mails, online chats, and browsing histories of millions of individuals. In one thirty-day period, there were forty-one billion records collected and stored by the program.[132] X-KEYSCORE even allows the analyst to intercept an individual's Internet activity live. The NSA visualizes its data collection through a program called Boundless Informant. By March 2013 the program showed the NSA had collected ninety-seven billion pieces of intelligence—much of it tied to IP addresses—from worldwide computer networks.[133] Controversially, Boundless Informant showed billions of U.S. records had been captured. Another tool for analyzing intercepted data is CO-TRAVELER. The NSA uses this program to geolocate individuals across the planet using billions of intercepted cell phone records.[134] The NSA can then employ powerful algorithms to retrace their movements, map social networks, and expose hidden relationships between people.

The ability to collect data about U.S. persons incidentally provides a potential back door, or loophole, for widespread warrantless surveillance of Americans.[135] Although U.S. citizens' identifying information is minimized, it can be unmasked at a later point in time. Furthermore, a danger exists this information could be used surreptitiously to drive criminal investigations under a controversial process known as "parallel construction." This "involves law enforcement agents using information gleaned from covert surveillance, but later covering up their use of that data by creating a new evidence trail that excludes it. This hides the true origin of the investigation from defense lawyers and, on occasion, prosecutors and judges—which means the legality of the evidence that triggered the investigation cannot be challenged in court."[136] The DEA has reportedly used this covert process, and as one former agent claimed, "It's just like laundering money—you work it backwards to make it clean."[137]

Beginning in 2007, the NSA constructed a Google-like search engine that allowed twenty-three U.S. domestic law enforcement agencies—including the FBI and the DEA—to trawl through 850 billion metadata records picked up in its dragnet.[138] The program is called ICREACH, and although the program is used for targeting foreign parties, it is likely that millions of records on U.S. citizens exist within the database. Although the NSA is a foreign signals intelligence agency, it constantly blurs the lines with domestic law enforcement. ICREACH, in particular, harks back to the days of SHAMROCK and MINERVA.

Government officials have endlessly reassured the public that it is only *metadata* for which the NSA typically hunts. But metadata can be extremely intrusive and easily give a picture of somebody's pattern of life. Moreover, it simply isn't true that surveillance captures metadata only. Under OPTIC NERVE the NSA used wiretaps from GCHQ to intercept nearly two million images from webcam chats.[139] The electronic program photographed a user once every five minutes, producing a complete profile of an individual: their e-mail history, their telephone calls, and their appearance. Facial-recognition technology could then be used to track the movements of individuals in their everyday life. Even online video game worlds have been infiltrated by NSA spies. In 2014 a UN special rapporteur wrote that mass surveillance amounts to "the systematic interference with the Internet privacy rights of a potentially unlimited number of innocent people in any part of the world."[140]

The NSA, Part III: The Killnet

The U.S. military has leaned heavily on the NSA for its counterterrorist operations in Iraq and Afghanistan—in particular, for its drone operations. The NSA created the Counter-Terrorism Mission Aligned Cell (CT MAC) to hunt down difficult terrorist targets. Soon after September 2001, the NSA assembled a team of analysts from the National Geospatial Intelligence Agency, the CIA, and the military to form the Geolocation Cell, whose motto was We Track 'Em, You Whack 'Em.[141] These NSA signals surveillance units were tactical cryptologic support teams and were housed at JSOC's Balad base in Iraq. By September 2004 the NSA was able to locate cell phones even when they were *switched off*. JSOC referred to the technology as the Find, and it produced thousands of new targets. Analysts could "track the cell phone to within 30 feet of its actual location, feeding the real-time data to teams of drone operators who conduct missile strikes or facilitate night raids."[142]

The geolocation system employed by the NSA was called GILGAMESH. The technology works by fooling cell phones on the ground to connect with NSA receivers attached to overhead Predator drones (which act as cellphone masts). As one JSOC drone operator said, "We're not going after people—we're going after their phones, in the hopes that the person on the other end of that missile is the bad guy."[143] A related system used to exfiltrate data is the NSA's SHENANIGANS program, which uses a similar technology. Specially modified Predator drones exfiltrate "massive amounts of data from any wireless routers, computers, smart phones or other electronic devices that are within range."[144] From miles up in the air, NSA analysts can capture the communications of people below. As one leaked document described the system, "Our mission (VICTORYDANCE) mapped the Wi-Fi fingerprint of nearly every major town in Yemen."[145]

The NSA is fully imbricated in the planetary manhunt, geolocating targets around the world and passing on its signals intelligence (SIGINT) to JSOC teams in countries that have ranged from Afghanistan to Somalia. Often depicted as a CIA manhunt, the drone strikes in Pakistan have relied heavily "on the NSA's ability to vacuum up enormous quantities of e-mail, phone calls and other fragments of signals intelligence, or SIGINT."[146] In many cases it is *only* SIGINT that drives the kill. One former JSOC drone operator said the vast majority of high-value target operations in Afghanistan relied on the Geolocation Cell. "Everything they turned into a kinetic strike or a night raid was almost 90 percent that. You could tell, because you'd go back to the mission reports and it will say 'this mission was triggered by SIGINT,' which means it was triggered by a geolocation cell."[147] Tracking by metadata and killing by SIM card are the purest expressions of the thanatopolitics pursued by the Predator Empire's electromagnetic enclosure of the planet.

Perhaps, the very idea of a military kill chain needs to be rethought, given the widespread, multidimensional, and multi-institutional flow of data that finds, fixes, and finishes a target in the global manhunt. Internet search histories, Wi-Fi data, computer keystrokes, cell phone conversations, Facebook posts, text messages, and e-mails are variously intercepted by satellites, drones, planes, computers, listening posts, worms, and cable taps belonging to the NRO, CIA, NSA, JSOC, and FBI. As such, *kill network,* or *killnet,* might be a more accurate term for describing the shape of contemporary targeted killing. While the idea of a killnet may risk exaggerating the resonance between data, analysis, and death, it lends a more accurate shape to the multiple, dispersed, and violent infrastructures that have snapped together in the Predator Empire.

Machinic Interpellation

The Predator Empire imposes a systematic rule by Nobody: a technical power generated by thousands of apparatuses that hack, surveil, and modulate the global lifeworld. "The new rules it brought into existence," writes Tom Engelhardt, "are simple enough: you—whoever you are and wherever you live on Planet Earth—are a potential target. Get used to it. The most basic ground rule of the new system: no one is exempt from surveillance."[148] With remarkable ease the Predator Empire listens as the world talks. As Snowden reflected, "I realized that they were building a system whose goal was the elimination of all privacy, globally. To make it so that no one could communicate electronically without the NSA being able to collect, store, and analyze the communication."[149] Ever restless, the Predator Empire spreads its wings as it roams across the planet, diving into oceans and minds alike. Humans have become unavoidable subjects of a pervasive machinic interpellation, as bodies and neurons are policed, reconfigured, and scanned by nonhuman policing apparatuses across the world.

The control society discussed by Deleuze has become the definitive biopolitical model for planetary policing under the Predator Empire. Enclosed by synthetic shells, a pernicious conformity chills the modern individual's ability to act and think independently. Machines continually unleash themselves on the planet, processing, manipulating, and redistributing the affective climate of our coexistence—producing atonic worlds that strip humanity of its need to think freely and act autonomously. For Stiegler "it is not simply a standardization process: it is an automation one, that is to say a massive transfer of the psychical, mental and social functions towards machines and industrial devices."[150] Machines police the phenomenological profile of the world, enclosing bodies, minds, and things inside secured atmospheres. As Jacques Rancière states,

"Policing is not so much the 'disciplining' of bodies as a rule governing their appearing, a configuration of *occupations* and the properties of the spaces where these occupations are distributed."[151] Machines must therefore be understood as *force-full*: thoroughly political technologies that remap the flow of physical and psychological forces across the spheres of technological civilization.

The enclosures of control societies are neither built with walls nor bound to concrete spaces. A defining feature of the Predator Empire is its systematic surveillance of the planet's population across amorphous technogeographies that undercut established notions of territory and sovereignty. "The NSA's global panopticon," argues Alfred McCoy, "thus fulfills an ancient dream of empire."[152] The endgame of this global panopticon is for everybody to think and act with a machinic predictability, thereby installing a perfect peace across the planet. If our neurons are already disciplined and our thoughts are in perfect acquiescence, mass surveillance has served its purpose: totalitarian synchrony. "Power," write Hardt and Negri, "is now exercised through machines that directly organize the brains (in communication systems, monitored activities, etc.) toward a state of autonomous alienation from the sense of life and the desire for creativity."[153]

George Orwell's *1984* describes the "essential crime" that a controlled population could commit: free thought. The NSA's mass surveillance represents an insidious assault on our minds, even if its infrastructures are not obviously targeted against domestic populations. The very potential that our emails *could* be read can be just as damaging to liberty. "In closed societies," writes Naomi Wolf, "this surveillance is cast as being about 'national security'; the true function is to keep citizens docile and inhibit their activism and dissent."[154] If we think we are being watched, we alter our behavior in both obvious and subtle ways. Privacy in such times is a privilege rapidly evaporating. As Greenwald argues, "Far more effectively than a police force, the deprivation of privacy will crush any temptation to deviate from rules and norms."[155] In 2014 Snowden revealed, "As you write a message, you know, an analyst at the NSA or any other service out there that's using this kind of attack against people can actually see you write sentences and then backspace over your mistakes and then change the words and then kind of pause and . . . think about what you wanted to say. . . . And it's this extraordinary intrusion not just into your communications, your finished messages but your actual drafting process, into the way you think."[156] Here, the target is not only what we say but how we think. Surveillance operates on this deeply psychological level, policing the thought patterns of millions. Surveillance is not, therefore, a passive force, as is often assumed: it is an active force of interpellation. The freedom to think in such times has never been more important. As Deleuze and Guattari warn, "The less people take thought seriously, the more they think in conformity with what the State wants."[157]

Although we currently associate the NSA's programs with a Democratic president (and for many this implies a degree of trust), the future is always uncertain. What counts as acceptable thinking, or as safe thinking, is continually narrowing. After discovering the U.S. government's panopticon in 1975, Senator Frank Church warned, "That capability at any time could be turned around on the American people, and no American would have any privacy left, such is the capability to monitor everything: telephone conversations, telegrams, it doesn't matter. There would be no place to hide."[158] The NSA's hardware, in other words, embodies what Church feared was a capacity "to make tyranny total in America."[159] Crucially, we cannot simply regulate our way out of this panopticon. Nor can we install a benevolent ghost in the machine. So long as the material infrastructures exist, the human condition suffers. As the planet's populations become more unequal, as more and more individuals find themselves surplus to economic demand—tossed aside by a technological civilization that functions with less and less human input—there is a real danger that ordinary citizens, American or otherwise, will find themselves caught in this dragnet for no other crime than their superfluity. And yet even after the crushing totalitarianisms of the twentieth century, Arendt still believed that in moments of moral collapse, when the world itself seemed lost, thinking may yet prevent catastrophe.[160]

5

Policing Everything

Policing Civilization

From outer space to the high seas and now to the suburbs, the policing of the U.S. homeland concludes the Predator Empire's full spectrum dominance of the planet. This trajectory, which has seen the police become ever more militarized, must be situated, however, within a more pervasive matrix of social war that grinds within technological civilization. Our understanding of contemporary policing needs to be set against the ongoing and historical practice of enclosure, which is productive of an inescapable civil conflict. Whether we analyze the war on drugs, the increased use of SWAT teams, the spread of robotic systems across the city, or even predictive policing, all of them emerge from—and seek to control—a foundational tear in the heart of society. So as the police begin to deploy swarms of microsized drones across a splintered and militarized urban landscape, we must listen for the rumble of a social war entombed deep in the concrete—for beneath our feet lies the fossilized injustices of centuries of enclosure. While the 2014 shooting of Michael Brown in Ferguson, Missouri, and the heavy-handed tactics that followed might viscerally expose the state of U.S. police militarization, this was an event born of prevailing conditions—namely, the great age of confinement.

Modern U.S. policing, unsurprisingly, was forged in the crucible of war. An important example of this came at the turn of the twentieth century when the U.S. military annexed the Philippines after a victory against the Spanish. During the four years it directly controlled the country, between 1898 and 1902, the military ran a pacification program against Filipino nationalists. Alfred McCoy argues the contemporary U.S. police state was built in this distant counterinsurgency.[1] The U.S. Army institutionalized advanced data-management techniques in the Manila Metropolitan Police. Within two decades this agency created an all-embracing system of alphabetized file cards for securing Manila's population. In turn, the geographies of military surveillance spiraled home: "Washington repatriated the personnel and policies of colonial rule during World War I and used them to conduct what may have been the most systematic surveillance of its citizens ever undertaken by a modern government, producing institutional innovations that helped establish a nascent national security state."[2] U.S. society was "honeycombed" with a matrix of "active informers, secretive surveillance organizations, and government counterintelligence agencies."[3] For example, during the Red Scare that followed World War I, U.S. military intelligence continued its covert campaign against labor unions, using tactics from the Philippines to smash a radical miners' revolt in the West Virginia coalfields. "Empire thus proved mutually transformative in ways that have arguably damaged democracy in both the Philippines and the United States."[4]

From the outset, then, the apparent division between the police and the military is complicated. The belief in separate military and police institutions masks their common function: to secure technological civilization. Under this renewed understanding, the police become more than an institution staffed by officers—they become an order or a law that crisscrosses the social landscape to arrange bodies, events, and thoughts. In a similar vein, Michael Hardt and Antonio Negri argue empire is immanent to social life: it "creates the very world it inhabits" and rules over "social life in its entirety."[5] Policing is a power of social pacification that restricts the possibilities of being-in-the-world by modulating human environments and creating artificial atmospheres of control. What Michel Foucault calls the "carceral network" has no "outside," and its systems of surveillance create a "normalizing power" across society.[6] By dismantling our commonsense assumptions about the police, we begin to see apparatuses of control everywhere.

Peace is typically assumed to be the natural condition of civilization—the blissful state for which humanity clambers in its escape from primeval struggle. But a great danger is posed by accepting this a priori formulation. If the civilizatory system is assumed to exist in a peaceful state, it cannot subsequently generate conflict from its own inner workings. Crime, terrorism, and the generalized existential condition of insecurity must consequently come from the *outside*. The police thereby become the system's protectors, triumphantly guarding civilization from the hell beyond the wall.

But we must turn this whole understanding on its head. As Foucault asks, "If we look beneath peace, order, wealth, and authority, beneath the calm order of subordinations, beneath the State and State apparatuses, beneath the laws, and so on, will we hear and discover a sort of primitive and permanent war?"[7] Sovereignty as such needs to be understood "in terms of the unending movement—which has no historical end—of the shifting relations that make some dominant over others."[8] The Leviathan is always gnawing at its own body, and the various state apparatuses have war as their political nucleus.

To oppose civilization with savagery, to oppose war with peace, is thus to make a fundamental misstep. The war on terror, for example, endlessly marshals the interrelated concepts of civilization and savagery. But as Mike Davis argues, "This is not a war of civilizations but an oblique clash between the American imperium and the labor-power it has expelled from the formal world economy."[9] Civilization therefore generates many of the very insecurities it seeks to secure. Peace is a constant war of pacification. To recall Mark Neocleous's argument, "The 'war on terror' is thus the violent fabrication of world order in exactly the way that the original police power was the violent fabrication of social order. The war on terror, as international ordering, is a form of police; civilisation writ large."[10]

From the outset this police order has been overdetermined by an economic logic. In England the model of enclosure was perfected and universalized by the spiritual maxim of technological civilization. England was certainly a paragon of this geosocial engineering, but the process was soon repeated across the planet, becoming a crucial colonial export and integrating into the world system. Peasants and the common land they inhabited were seen as barbaric and backward.[11] In field after field, peasants were thrown off the soil like inconvenient wildlife, impediments to the divine progress of the economic system. English society was destroyed by capitalist property definitions, and masses of people became "strangers in their own land," to use E. P. Thompson's famous phrase.[12] Landlords believed the commons needed social as well as economic improvement. As Neocleous explains, "The belief was that if left unimproved the wasted commons would generate a masterless, idle and disorderly mass."[13]

Thus, an *internal* colonialism took root long before 1492, a kind of pacification of the commons. Under such an understanding, the military and police rather than being exclusionary forces are both arms of state power, utterly invested in "the most fundamental and violent conflict in human history: the class war."[14] By excavating this fiery history, the fractured ground of technological civilization is revealed. Thousands of years before drones spilled into the sky, a series of existential conditions were dividing the planet's inhabitants. In the age of electronic globalization, this social war has spread to become a planetary condition, and policing has become imperial policing. Everyone and everything must be kept in their place, peacefully locked inside the civilizatory enclosures. Today, as technological civilization becomes ever more capital intensive, as more and more humans are displaced by robotic systems, the spiraling contradictions are policed with ever greater force.

Setting in Motion the Great Age of Confinement

As discussed, the history of English enclosure is important in understanding how humanity slowly succumbed to the great age of confinement. Conducted over centuries, this form of widespread agricultural enclosure created masses of newly "liberated" wage laborers. In turn, this surplus population generated a policing problem for the state—one that remains at the heart of a splintering technological civilization.

Karl Marx famously labels the original sin of capitalism as enclosure, or what he calls "primitive accumulation."[15] This is "the historical process of divorcing the producer from the means of production."[16] In other words, it is the severance of humans and earth: *the originary act of worldly alienation.* Before primitive accumulation, there was not the division between natural resources and labor we imagine today. Undoubtedly, the commons—that is, commonly held or accessed land—were lively and at times acrimonious places. But nonetheless, a durable coexistence prevailed: a more seamless ecosystem of people, things, and animals. For this reason Peter Linebaugh argues the commons is best understood as a verb, *commoning.*[17] By enclosing this ecosystem, people were robbed of their ability to survive autonomously. "And the history of this, their expropriation, is written in the annals of mankind in letters of blood and fire."[18]

By the fifteenth century the English population consisted largely of rural peasants, many of whom inhabited land under some form of customary right, which typically allocated them access to commonly held land. Legal documents such as the 1100 Charter of Liberties, the 1215 Magna Carta, and its companion document, the 1217 Charter of the Forest, all went some way toward protecting these customary rights.[19] More often, the laws of the commons were simply unwritten and existed in the everyday habits and practices of commoners.[20] But wealthy landowners and politicians viewed this social geography as an impediment to national economic progress, and so the commons soon became a hunting ground for a predatory form of capitalism. Enclosing common land and smaller peasant holdings into exclusive manorial property was an early method used to generate capital for landlords. As time ticked away, more systematic forms of enclosure emerged. In the sixteenth century, for example, the breakdown of feudal retainers was mandated by royal prerogative under the reigns of Henry VII and Henry VIII. The latter king was also instrumental in dissolving church property under the Reformation. This land was often vital to the sustenance of poor people. The most systematic assault on the commons came with parliamentary enclosure.

The first such enclosure act was passed in 1604, and legislation expropriating land from commoners accelerated over the next two hundred years, reaching its peak with the 1801 General Inclosure Act. Between 1604 and 1914, over 5,200 enclosure bills were enacted in Parliament, relating to an area one-fifth the size of England, or some 6.8 million acres.[21] By the end of the nineteenth century, open fields were fenced, and many of the natural waterways of the world were privatized: rivers were canalized, and port traffic was enclosed by docks.[22] In other words, the various English enclosure acts were enforced by modifications to the landscape: ditches, hedges, and walls divided and separated the commoners, generating a civil conflict that unavoidably was a type of geographical war. Crucially, this geoengineering created a widespread existential shock for the people that once dwelt there. As Marx observes, lords "created an incomparably larger proletariat by the forcible driving of the peasantry from the land, to which the latter had the same feudal right as the lord himself."[23] Independent peasants would now have to buy back the common treasures of humanity.

Resistance against the imposition of this growing transcendental system was widespread. Commoners were not passive victims: they derided the notion that customary rights attached to a place could be transferred to *individuals*. "Riotous resistance," even arson, greeted the early enclosures at the turn of the fifteenth century.[24] "The more that one looks, the more that one finds such disputes to be normal, in great towns and in small."[25] Nonetheless, usage rights were slowly traded for property titles, and the stability of the world was uprooted and set adrift. "First, in the process of primitive accumulation," write Hardt and Negri, "capital separates populations from specifically coded territories and sets them in motion."[26] All that was solid melted into a deluge of social flotsam. Families and villages were shipwrecked on dry land, and the coexistence between humans and earth was refashioned into something altogether alien.

With the ascent of absolute notions of property, the toleration for practices such as open pasture and wood collecting ended. Peasant rights were usurped by property rights, and commoning was rebranded as theft. Ancient habits were vanquished by a wealthy cabal that recognized only the ink spilled on government paper. Politicians, landowners, and agricultural boards were in agreement. "They were so profoundly imbued with preconceptions which translated the usages of the poor into the property-rights of the landowners that they really found it difficult to view the matter in any other way."[27] A strong government became necessary to mediate and manage this expropriation. As Linebaugh argues, "The state plays a decisive role in enclosure: its servants and warriors write the letters of blood and fire."[28] Parliament, from the outset, was an arbiter of the contradictions thrown up by the Industrial Revolution.

Enclosure shifted the terrain of what activities were defined illegal. As Foucault explains, "The transition to an intensive agriculture exercised, over the rights to use common lands, over various tolerated practices, over small accepted illegalities, a more and more restrictive pressure."[29] This distribution of illegalities was structured so that lawmakers—the landowners—created a legal system favorable to themselves. There was nothing, therefore, particularly just about parliamentary enclosure. As Thompson famously writes, enclosure "was a plain enough case of class robbery, played according to fair rules of property and law laid down by a parliament of property-owners and lawyers."[30] So while enclosure may have been a lawful act of Parliament, it was by no means just. Parliament simply legitimated the unequal relationship between landowners and a landless working poor. "If it is pretended that the law was impartial, deriving its rules from its own self-extrapolating logic, then we must reply that this pretence was class fraud."[31]

In a matter of decades, a profound social war had been codified into English law. Enclosure was a project of geoengineering, a psychosocial shock, and above all a form of war: "*the law of private property as war.*"[32] Enclosure was thus a perfectly legal form of social war. Recall Foucault, who argues that law "is born of real battles, victories, massacres and conquest."[33] Friedrich Engels similarly writes capitalist society is a "war of each against all."[34] Marx calls it a "civil war in its most terrible aspect, the war of labour against capital."[35] He continues by stating the robbery of common lands was a case of "ruthless terrorism" that "created for the urban industries the necessary supplies of free and rightless proletarians."[36] This free and rightless proletariat threw a giant wrench in the emerging industrial system. How could capitalism flourish in a world birthed by dispossession, riot, and surplus populations? Enclosure, as Neocleous observes, thus created "a fundamental police problem: how to generate a peaceful and secure order of lawful obedience amid the apparent disorder and insecurity of bourgeois society."[37]

The exiled masses routed by enclosure now found themselves "turned in massive quantities into beggars, robbers, vagabonds."[38] The accumulation of capital generated an accumulation of worldless people. In the sixteenth century a raft of draconian laws against vagabondage was written. Under Henry VIII, for example, vagabonds could be whipped, disfigured, imprisoned, and executed. "Thus were the agricultural folk first forcibly expropriated from the soil, driven from their homes, turned into vagabonds, and then whipped, branded and tortured by grotesquely terroristic laws into accepting the discipline necessary for the system of wage-labour."[39] With the ascent of the Industrial Revolution, the choice was clear: newly "liberated" proletarians could either work for a wage or turn to a life of vagrancy. A divided world was being engineered by enclosure, and an industrializing civilization had to forcibly subdue its alienated denizens. "This is the meaning of police," writes Neocleous, "that eventually becomes central to its whole history and logic: the fabrication of *durably pacified social spaces.*"[40]

It is little wonder that as enclosure reached its fierce crescendo in the nineteenth century a new form of social control was urgently needed. The theatrical display of power associated with torturing and hanging criminals in public squares was ineffective against an endemic social war that drew in the wider population. Sovereign power had to be redistributed away from kings and queens into the atomic relations of society in order to secure the newly installed enclosures. As Foucault writes, "There emerged the need for a constant policing concerned essentially with this illegality of property."[41] After masses of exiled people were forced into squalid urban environments, an exploding crime rate led Parliament to create the urban constable. A key date in this genesis is 1829, when Sir Robert Peel created the Metropolitan Police Force in London.

A form of systematic discipline was emerging to pacify capitalism's restive population, "a power to punish that ran the whole length of the social network."[42] The landscape had already been geo-engineered with fences and hedges. Now, the laboring poor had to be disciplined themselves. A raft of what Foucault calls "disciplinary institutions" began to spring up across the country. "Discipline," writes Foucault, "sometimes requires *enclosure,* the specification of a place heterogeneous to all others and closed in upon itself. It is the protected place of disciplinary monotony."[43] So as the commons emptied, the prisons began to fill—two sides of the same coin. As Linebaugh observes, "A massive prison construction program accompanies the enclosure of agricultural production."[44]

All of the disciplinary institutions—prisons, factories, hospitals, schools, workhouses, and asylums—had as their institutional nucleus the model of social conflict. "Is it surprising," asks Foucault, "that prisons resemble factories, schools, barracks, hospitals, which all resemble prisons?"[45] Hundreds and thousands of state-sanctioned enclosures began sticking dense masses of wandering and maladjusted people together, fixing them inside the same place. The tragic paradox can be summarized as follows: *state enclosures were built to mitigate the effects of state enclosure.*

Parliament's reaction against the violent contradictions of enclosure was to create a society full of enclosures. These disciplinary spaces began to regulate what was normal in every facet of life. The school and its teachers, the army and its generals, the asylum and its doctors, the factory and its overseers all instilled discipline upon a potentially restive and worldless population. The great age of confinement and surveillance was installed upon the ashes of communal living. Across much of Europe, a profound architecture of control was redistributing, separating, and atomizing individuals across space. These enclosures circulated a microphysics of power in the population that sought to reform deviant souls and produce docile bodies. The "dream of a perfect society," argues Foucault, belongs not only to the philosophers of the eighteenth century but also to the military, whose "fundamental reference was not to the state of nature, but to the meticulously subordinated cogs of a machine."[46]

State enclosures policed space in highly precise and methodical ways to fix and partition deviant and surplus mobilities. This spatial distribution of bodies is apparent in prisons, schools, hospitals, and factories. "In organizing 'cells,' 'places' and 'ranks,' the disciplines create complex spaces that are at once architectural, functional and hierarchical. It is spaces that provide fixed positions and permit circulation; they carve out individual segments and establish operational links; they mark places and indicate values; they guarantee the obedience of individuals."[47] It was of course in the modern prison that Foucault uncovered the most distilled form of discipline. Jeremy Bentham's widely discussed Panopticon remains an architectural paragon.

Bentham designed a circular prison with cells that ringed a central tower. "The theme of the Panopticon—at once surveillance and observation, security and knowledge, individualization and totalization, isolation and transparency—found in the prison its privileged locus of realization."[48] Inmates could be watched at any time without knowing whether they were being observed. The Panopticon ensured the gaze was one way, installing a permanent and unilateral visibility. As Foucault argues, "This architectural apparatus should be a machine for creating and sustaining a power relation independent of the person who exercises it."[49] The Panopticon was never built, but it birthed a political technology, a blueprint, that could be endlessly replicated in space and time.

Discipline may have begun in claustrophobic state enclosures, but it soon spilled out from these quarantines to enclose society under a generalizable mechanism of panopticism.[50] These pacified institutional spaces produced a universal economy of behavior, a generalizable model of society. "As the walls of these institutions break down," argue Hardt and Negri, "the logics of subjectification that previously operated within their limited spaces now spread out, generalized across the social field."[51] Enclosure went further by becoming smaller. Instead of a single Big Brother, a unified global Panopticon, "we have a multitude of Little Brothers—an 'omnopticon' encompassing multiple surveillance systems of diverse scope, scale, effectiveness and reach."[52] Carceral forms of enclosure were thus replicated in smaller and more intimate sites of social activity.

Technological civilization was enclosing and devouring its subjects, placing them inside an archipelago of enclosures that resonated together to form an emerging open prison. Human coexistence had been fundamentally reengineered on planet Earth. As Linebaugh concludes, "The incessant accumulation of 'industrial' subjects required their enclosure from the cradle to grave. To be ruled the population of civil society had to be confined and to be confined it had to be brought under complete surveillance."[53]

The birth of the Industrial Revolution and the concentration of human beings within stifling urban densities meant the state's chief concern soon became managing and pacifying the city. Policing a restive urban landscape required a responsive disciplinary architecture, one that would increasingly come to rely on nonhuman forms of mediation. The "urban problem," Foucault writes, is centered on "control over relations between the human race, or human beings insofar as they are a species, insofar as they are living beings, and their environment, the milieu in which they live."[54]

The whole problematic of enclosure thus becomes a question of pacifying a growing urban population. How could the state concretize a widespread disciplinary function within the city itself? A classic response comes from Paris. Under the revolutionary throes of mid-nineteenth-century France, Louis-Napoléon Bonaparte hired Georges-Eugène Haussmann to oversee an enormous modernization of Paris. The construction of airy boulevards and shopping malls reengineered Paris as a city of light. Crucially, it physically removed the narrow and dark medieval streets that were cradles of popular insurrection. Cafés and open streets thus served to pacify the population. Haussmann built a city in which, to quote David Harvey, "sufficient levels of surveillance and military control could be attained to ensure that revolutionary movements would easily be brought to heel."[55] As what Harvey calls the "embourgeoisement" of Paris reveals, the urban landscape is always a type of political landscape. Matter is mutable: it can be reshaped by the political forces that batter and bruise it. These ensuing forms subsequently police the social landscape. As Harvey concludes, "Urbanization has always been, therefore, a class phenomenon, since surpluses are extracted from somewhere and from somebody, while the control over their disbursement typically lies in a few hands."[56]

This expression of urban power is an objective—rather than subjective—form of power. It constellates a pattern in the planet's concrete, etching power relations into the ground. Urban power doesn't necessarily rely on police patrolling a beat but can be realized in the city's material furniture. A small but telling recent example is the antihomeless metal studs now appearing in Western cities. These spikes are fused into the ground to prevent vagrant populations from sleeping in high-rent districts. This hostile architecture, or revanchist urbanism, is a punitive ordering of space based on the requirements of capitalism rather than on human needs. The neoliberal state roots out the enemies of economic order as it polices the uneven contours of gentrification—another polite term for an ongoing form of social war. The so-called broken windows policing enacted by New York City mayor Rudy Giuliani is an important example. This model of zero-tolerance urbanism has been endlessly exported. "The revanchist city," writes Neil Smith, "stretches well beyond New York, and the criminalization of more and more aspects of the everyday life of homeless people is increasingly pervasive."[57] From hedge to prison to metal stud, the technologies of enclosure may have transformed, but their function—to police, to subdue, to pacify—persists in the great age of confinement. Indeed, "we must now add the bomb and the drone as a means of expropriation."[58]

The material landscape is fully invested in the war of enclosure that traverses society. The relationship between forces and forms, a type of endless loop, is crucial to this understanding. Forces create material forms, which in turn generate further forces. Politics is always already materialized and materializing: it is a mutable and plastic force that opens and encloses the world. Eyal Weizman uses the wall that divides Israeli and Palestinian territories to make this point. He argues the wall solidifies "the material imprint of forces."[59] But this architecture is not simply a passive recording device. The wall is active in the political and social order. As Weizman continues, "Space is not a representation of a politics that would already otherwise exist in the abstract. Politics operates and flows *through* and *in* spatial practice."[60]

Rather than being divorced from things and the terrain, politics—which is to say *geopolitics*—is actively distributed in and through their spatial arrangement. Indeed, territory is always the outcome of this shifting arrangement between terrain and architecture. As Steve Herbert explains, "The processes of internal pacification so central to the authority of the modern state readily depend on the capacity of the police to mark and enact meaningful boundaries, to restrict people's capacity to act by regulating their movements in space."[61] Thus, instead of conceptualizing space as a neutral background to power, or as the product of political forces, Weizman argues we "have to imagine a different consistency of space"—that is, "a common plane, gelatin like, on which those forces are simultaneously existing and interacting."[62] Whether nation-state, battlefield, or city street, state power is maintained by diverse sociotechnical infrastructures and their attending technogeographies.[63] More than ever, these transduce space with computational forms of control.

Today, as discussed in chapter 4, a principal method for policing the city of the control society is the embedding of software code into its surfaces. Here, digital systems transduce space: opening, closing, and partioning the city according to binary codes.[64] These transductive materials police the circulations of life more totally than do the inanimate walls of disciplinary society, organizing the distribution of life with precise electronic calculations. After hundreds of years, the architectures of enclosure have certainly advanced from their organic beginnings, even if the desire to manufacture terrains comprised of pacified social spaces remains. This means the technologies the police use actively reshape the constellation of law enforcement, space, and the population.

U.S. Domestic Counterinsurgency

As enclosure was tearing through England during the eighteenth century, building the spaces for the great age of confinement, British soldiers across the Atlantic were busy creating another legacy. Long before the Filipino counterinsurgency, early American policing was shaped by colonial contact with the British. Without evidence of criminal activity, British soldiers in New England regularly searched homes under a much despised general warrant known as a "writ of assistance." Under the Quartering Acts of 1765 and 1774, colonists were further required to house and feed British soldiers. This came despite Britain's aversion to the quartering of soldiers in its own towns and cities—a practice banned under the English Bill of Rights in 1689. As Radley Balko explains, "Bostonians were British subjects, but they were being treated like enemies of the state."[65] The early founders were thus profoundly aware of the toxicity of militarism. The aversion to quartering and general warrants was later enshrined in the Third and Fourth Amendments to the U.S Constitution. At the heart of the entire edifice was the Castle Doctrine, which held an individual's home was sacred: police could not enter without a court warrant and must announce themselves by knocking on the door.

But how long could memory of the British experience serve as a buttress against government overreach in the new republic? For much of its early history, the United States respected the firewall between the police and the military. There were ebbs and flows, of course. The 1792 Militia Acts and the 1807 Insurrection Act permitted the president to call up militia in response to lawlessness and rebellion. After the Civil War, for example, federal agents were used to enforce the Reconstruction Acts from 1870 onward in the Confederate South. The acts—which formally prohibited slavery—were difficult to impose in the slave-owning South, and federal troops helped contain a backlash of mob violence and lynching. Congress later passed the 1878 Posse Comitatus Act to prevent domestic law enforcement officials from using troops to bolster their power. Posse comitatus has since become a symbol against U.S. militarism, even if its roots suggest it was partly a result of sympathy for the Confederacy.[66]

The twentieth century ushered in a terminal decline in the great American firewall. At its dawn the Militia Act of 1903 was passed, replacing the 1792 Militia Acts and creating the modern National Guard. Significant to this history is the event of Little Rock, Arkansas, in 1957. After the Supreme Court's decision in *Brown v. Board of Education* to end racial segregation in schools, nine black students attempted to attend class at Central High School, where a mob awaited their arrival. But it wasn't just angry southerners standing at the front doors. Governor Orval Faubus had blocked the students' entrance with National Guard troops. In response, President Eisenhower federalized the Arkansas National Guard and ordered soldiers from the 101st Airborne Division to escort the students. Little Rock is an infamous (and rare) example of direct military intervention in U.S. society. But the twentieth century witnessed something far more insidious. The 1960s civil rights movement unleashed the *indirect* militarization of U.S. policing. Cops were slowly becoming more like soldiers, something the founders hadn't predicted.

"A riot," proclaimed Martin Luther King Jr., "is the language of the unheard." Riots erupt along the faultline of an entrenched social war. Years of animosity between the Los Angeles Police Department (LAPD) and the city's black population underscored the Watts riots of 1965. Amid violent scenes of unrest, one thousand people were injured and thirty-four were killed. White America watched in horror as a "criminal class" took to the streets. In many ways, this social war was a long time coming, and not just because of endemic sociospatial apartheid. At the beginning of the century, modern U.S. policing began to professionalize. After widespread corruption, efforts were made to rationalize U.S. policing through bureaucratic management, moving policing away from the influence of city and machine politics. "There was of course an important downside to this new force," writes Herbert, "its detachment from the community made it less responsive than many political leaders wished to the ongoing complaints about racist uses of excessive force."[67] Chief William Parker, who served between 1949 and 1966, was an advocate of this style of policing and first spearheaded the professionalization of the LAPD in the 1950s. The force was transformed into a more sealed, hierarchical, and authoritarian agency, with "cops indifferent to the areas they patrolled" and inculcated to believe "they were all that stood between order and anarchy."[68]

In this context the seeds of a racialized battleground were sewn. As Herbert argues, "The image of the Los Angeles Police Department as an agency designed to protect white citizens from the influx of dark-skinned immigrants was avidly embraced by Chief Parker during the 1950s and 1960s."[69] This philosophy was shared by Parker's protégé and successor, Chief Daryl Gates, who served from 1979 to 1992. If Parker believed that policing was a moral crusade against "wicked men with evil hearts who sustain themselves by preying upon society,"[70] then Gates thought no differently. "Society," he claimed, "flinches from the truth: we do our very best to find psychological and sociological reasons to excuse behavior that our minds won't accept for what it is. You walk into court and you have all these attorneys explaining away all of the things that you can sum up in one simple world: Evil."[71] To oversee the 1965 Watts riots, Parker hired Gates, who later called in 13,500 California National Guard troops to reinforce the LAPD.[72] But Gates's legacy extends beyond Watts. In 1969 he created the Special Weapons and Tactics (SWAT) police unit. The first-ever SWAT raid was on the Los Angeles headquarters of the Black Panthers. And it was a disaster. But no matter, it set in motion a ruinous lurch toward a machismo-infused military policing in the United States.

The war on drugs provided the vehicle for the growing swatification of American policing. Richard Nixon seized the opportunity presented by civil rioting to launch a new social war in America. Nixon's war on drugs was baptized in a June 17, 1971, speech, in which he declared, "Narcotics addiction is a problem which afflicts both the body and the soul of America" and that the government must "wage an effective war against heroin addiction."[73] Only a month earlier, a Nixon-appointed committee came to the opposite conclusion. It identified the root causes of drug addiction as "deep societal ills" that "increase the individual's sense of personal alienation."[74] But the administration wasn't interested. As Emily Dufton argues, "Once addicts were no longer seen as sick victims of a society that systematically excluded them, no one would mind when they were simply locked up. In fact, incarceration was for the nation's *own good*."[75] So began the moral crusade. A growing list of U.S. states began to adopt Nixon-style antidrug legislation to pacify what was perceived as a deviant population. But militarism hadn't quite strangled policing yet. That coup d'état belonged to a future president.

Ronald Reagan's "drug warriors were about to take aim at *posse comitatus*, utterly dehumanize drug users, cast the drug fight as a biblical struggle between good and evil, and in the process turn the country's drug cops into holy soldiers."[76] Reagan oversaw the 1981 Military Cooperation with Civilian Law Enforcement Agencies Act, which paved the way for future combined military–police operations, particularly in counternarcotics. In 1986 he signed National Security Decision Directive 221, which named the drug trade a national security threat, thus inviting the military to police the global trafficking of narcotics. Reagan also permitted police to confiscate criminal assets and use them to fund SWAT teams, creating a precedent for today's controversial civil forfeiture programs. With federal grants like the Justice Department's Byrne Grant Program, cash was surging into SWAT teams. Perhaps, the final nail in the coffin came in 1987 when the secretary of defense and the U.S. attorney general were required to notify law enforcement about surplus military equipment.[77] Congress had engineered a system in which weapons from the battlefield could be transferred to "American streets, in American neighborhoods, against American citizens."[78] By the close of the decade, nearly every large U.S. city had a SWAT team on standby.

Reagan's sentencing reforms—in particular, mandatory prison sentences—fueled an explosion in the incarcerated population.

Complicated socioeconomic issues were simply thrown behind bars. African Americans were disproportionately targeted. Today, the system remains deeply sick. Between 2011 and 2012, for example, 62 percent of SWAT deployments were for drug searches—not the emergency scenarios envisaged in the 1960s.[79] The United States spends over $51 billion a year on the battle against narcotics users, and in 2013 alone, 1.5 million people were arrested on nonviolent drug charges. This, in turn, feeds a prison–industrial complex that in 2013 incarcerated 1 in every 110 Americans, the highest rate in the world.[80] The war on drugs is directed too often against a marginalized, alienated, and abandoned segment of society. It aims to maintain the veneer of peace over deep straits of inequality, mental health issues, and sheer desperation. "To put it another way," writes Foucault, "we have to interpret the war that is going on beneath peace; peace itself is a coded war. We are therefore at war with one another; a battlefront runs through the whole of society, continuously and permanently, and it is this battlefront that puts us all on one side or the other."[81]

The 1990s continued the militarization of the police. Building on the 1987 precedent, in 1990 the 101st Congress passed the National Defense Authorization Act (NDAA). Section 1208 of this bill, "Transfer of Excess Personal Property," formalized the transfer of excess military gear to law enforcement for counterdrug activities. This act was broadened in the 1997 NDAA under section 1033, which created what is widely known as the 1033 Program. The hardware transferred under this program ranges from grenade launchers to armored personnel carriers.[82] Over eight thousand federal and state law enforcement agencies participate in the 1033 Program, which is overseen by the Law Enforcement Support Office, whose motto boasts From Warfighter to Crimefighter. Over $5.4 billion in gear has been transferred to U.S. police since the program started.[83]

The 1992 Rodney King uprising in Los Angeles—precipitated by the acquittal of the LAPD cops who were videotaped beating Rodney King a year earlier—saw two thousand people injured, fifty-three killed, and property damage in excess of a billion dollars. The outnumbered and overwhelmed LAPD were yet again reinforced by 13,500 troops from the California National Guard, the Third Battalion First Marine, and the Fortieth and Seventh Infantry Divisions of the U.S. Army.[84] What had effectively become an urban war zone—with tens of thousands of angry citizens taking to the streets—

could not be broken until the National Guard showed up. Indeed, the National Guard came to play an increasingly prominent role in U.S. policing: in 1992 alone it assisted in 20,000 arrests, searched 120,000 vehicles, and entered 1,200 private buildings without a search warrant.[85] The militarization of policing in the 1990s concluded with the Battle for Seattle in 1999. Peaceful protestors came under fire as Seattle law enforcement hurled pepper spray, tear gas, and stun grenades and even shot rubber bullets at Americans exercising their First Amendment rights. In short, throughout the twentieth century, U.S. streets were enclosed by a zero-tolerance model of policing.

Twenty-first-century militarized policing was galvanized after the terrorist attacks of September 11, 2001. The gargantuan Department of Homeland Security (DHS) provided new incentives for swatifying police departments with counterterrorism grants. When combined with the Byrne Grant Program, in 2013 alone the Department of Justice and DHS programs gave nearly $1.5 billion to state and local police departments for military gear and programs.[86] "Ultimately," writes Glenn Greenwald, "police militarization is part of a broader and truly dangerous trend: the importation of War on Terror tactics from foreign war zones onto American soil."[87] Tens of thousands of SWAT raids take place ever year, and the U.S. employs 120,000 federal law enforcement officers.[88] Given these immense financial incentives, it is unsurprising certain segments of the civilian population are reimagined as enemies and the street as a battlefield. As one police trainer writes, "We trainers have spent the past decade trying to ingrain in our students the concept that the American police officer works a battlefield every day he patrols his sector."[89] As long as there is a war out there—against crime, drugs, or terrorism—American policing will keep on manufacturing soldiers.

The U.S.–Mexico border is a particularly pronounced site where militarism and domestic policing collide. Indeed, the border is not a thin strip of territory—it is an extraconstitutional zone that extends one hundred miles inland. Billions of dollars in high-tech surveillance equipment is regularly flung along its two-thousand-mile corridor. The militarization of this border began in the 1970s and intensified after the 1994 launch of Operation Gatekeeper in San Diego.[90] Narcotics smuggling and undocumented crossings are frequently framed as a part of an ongoing border war. Between 1989 and 2012, the border patrol's budget increased 750 percent from

$232 million to $3.6 billion, and the number of agents has increased ninefold since 1998, reaching 21,444 agents by the end of 2011.[91] As Todd Miller and Gabriel Schivone argue, "Like the Gaza Strip for the Israelis, the U.S. borderlands, dubbed a 'constitution-free zone' by the ACLU [American Civil Liberties Union], are becoming a vast open-air laboratory for tech companies. There, almost any form of surveillance and 'security' can be developed, tested, and showcased, as if in a militarized shopping mall, for other nations across the planet to consider."[92]

The incendiary rhetoric and explosion in funding have legitimized the presence of U.S. troops at the border. For example, Joint Task Force North, part of the U.S. military's Northern Command, has supported border patrol agents in Arizona. Of course, federal legislation—principally the Posse Comitatus Act—should prevent this intermingling. To work around this law, the army supports the border patrol indirectly. One Tucson-based major explained, "The Soldiers used their state-of-the-art surveillance equipment to identify and report the suspected illegal activities they observed and vectored border patrol agents in to make the arrests and drug seizures."[93] The 650-plus miles of snaking concrete walls along the U.S.–Mexico border are an obvious manifestation of national enclosure. As Geoff Boyce and Jill Williams argue, "The proliferation of border fencing globally is a re-investment in the fortification of geopolitical boundaries as a buffer against threats narrated as simultaneously transnational and existential."[94] Beyond the wall other apparatuses of enclosure include the spectacularly inept Boeing virtual fence. Motion sensors—which first made their debut in the wake of Vietnam's Operation Igloo White—now number some twelve thousand across the desert.[95] The electronic battlefield is, in short, a permanent condition in the borderlands.

Warrior Cops in Ferguson, Missouri

On August 18, 2014, Missouri governor Jay Nixon signed Executive Order 14-09, calling into service an "organized militia" to help quell "civil unrest occurring in the City of Ferguson." Mine-resistant vehicles and camouflaged police officers occupied the city's road intersections, and the sky was declared a no-fly zone. By then the image of National Guard troops patrolling U.S. streets with military gear was no rarity, but the 2014 clashes still managed to capture the world's attention. On August 9, 2014, an unarmed black teenager named Michael Brown was shot six times by police officer Darren Wilson (who was later cleared by a grand jury). The shooting stoked fierce emotional responses from a black community that was policed by a predominantly white force. One black female police officer stated Ferguson residents existed in a "fear-based society."[96] Others labeled Ferguson a war zone, and the name Fergustan, after Afghanistan, circulated in social media. The idea that Ferguson had become a battlefield, however, masks the uneven exercise of power. Protestors chanted, "Hands up, don't shoot!" while police ringed them with assault rifles and fired tear gas into crowds. Some police reportedly arrived in trucks mounted with acoustic riot-control devices.

The 1033 Program was the object of renewed attention and criticism. Between 2006 and 2014, 93,763 machine guns, along with 533 planes and helicopters, were distributed to police.[97] The 1033 Program was briefly halted in 2013 after misguided appropriations made the headlines. Maricopa County in Arizona, for example, purchased a tank with a 360-degree rotating machine-gun turret that spat out .50-caliber bullets. It was called the Peacemaker. Between 2003 and 2012, St. Louis County law enforcement, which covers Ferguson, received $81 million in Urban Areas Security Initiative funds.[98] Officially, the St. Louis area received only night-vision sights, trucks, a bomb disposal robot, rifles, and pistols as part of the 1033 Program.[99] But police in Ferguson were photographed in armored cars known as Mine-Resistant Ambush Protected (MRAP) vehicles. The Pentagon spent $50 billion to produce 27,000 MRAPs for deployment in Afghanistan and Iraq.[100] As these wars ebbed, the blast-resistant vehicles flowed back to an estimated five hundred law enforcement agencies.[101] As Greenwald observes, "Americans are now so accustomed to seeing police officers decked in camouflage and Robocop-style costumes, riding in armored vehicles and carrying automatic weapons first introduced during the U.S. occupation of Baghdad, that it has become normalized."[102] The problem with military hardware, as Balko reflects, is that "when you're carrying a hammer, everything looks like a nail."[103] But the issue is bigger than hammers looking for nails, despite the importance of technology in mediating the conduct of state power.

Perhaps, then, the small army that occupied Ferguson was the direct result of decades of empire abroad. "When the war machine runs out of places to occupy abroad," writes Gilbert Mercier, "it mutates into an occupying force at home, starting in Black or Latino neighborhoods, and it manifests itself as police violence, curfews and a state of emergency."[104] Perhaps, however, this misses a more fundamental point. As Tamara Nopper and Mariame Kaba argue, "For blacks, the 'war on terror' hasn't 'come home.' It's always been here."[105] Indeed, as it turns out, a damning U.S. Department of Justice report reveals Ferguson's police department and court system reflected and exacerbated existing racial biases. As the report, which was released in 2015 after a six-month inquiry, states, "Over time, Ferguson's police and municipal court practices have sown deep mistrust between parts of the community and the police department, undermining law enforcement legitimacy among African Americans in particular."[106] In this respect the law enforcement structure in Ferguson was toxic from the very top. A systematic focus on revenue generation contributed to a pattern of aggressive enforcement and unconstitutional policing. Consequently, "many officers appear to see some residents, especially those who live in Ferguson's predominantly African-American neighborhoods, less as constituents to be protected than as potential offenders and sources of revenue."[107]

Although understanding the militarization of U.S. policing is important, it must be understood alongside the rampant inequality and racism across many of America's streets. As the FBI director himself admitted in 2015, racism is part of the U.S. "cultural inheritance."[108] In the case of Ferguson, the 1033 Program—and other programs like it—weaponizes an already existing social war. The robocops of Ferguson are just the latest warriors in a profound domestic conflict. In 2013, for example, the FBI counted 461 "justifiable homicides"[109] committed by U.S. police (a 2015 study suggests the real number is twice that).[110] This figure cannot be linked to the 1033 Program alone. Indeed, the United States has a long history of aggressively policing black communities. "It used to be billy clubs, fire hoses and snarling German shepherds," writes Kara Danksy. "Now it's armored personnel carriers and flash-bang grenades. The weaponry has changed, but the target is still the same."[111] By overly focusing on police militarization, the wider system it rests upon is masked. In 2014 and 2015, a string of accusations of police brutality against black men gripped the U.S. media. The real terror is thus far more pervasive and everyday than spectacular like Ferguson. Entire segments of society have been abandoned, oppressed, and enrolled in a social war. This is the fractured ground of technological civilization.

Future Policing, Policing the Future

The technological and biometric aspects of U.S. policing continue to develop in the Predator Empire—in particular, the practice of preemptive policing. Over the past two decades, U.S. policing has imported a raft of militarized techniques for dominating the electromagnetic spectrum. As societies have evolved from a model of discipline (with discrete enclosures) to one of control (in which discipline is modulated across the entire social field), the methods of enclosure have shifted, becoming increasingly computational and algorithmic. One of the most obvious manifestations of this change has been the growing use of electronic watchlists to monitor U.S. citizens, such as the SAR (suspicious activity report) database. The Nationwide Suspicious Activity Reporting Initiative (NSI) began its life in 2007 as a way for diverse public law enforcement agencies to fuse intelligence on domestic terrorism. According to a 2013 Government Accountability Office report, there were seventy-eight NSI fusion centers and 27,855 SARs by the end of 2012.[112] The system is designed to predict and preempt would-be terrorists by monitoring the population for suspicious activity. The SARs can amount, however, to invasions of privacy protected by the First Amendment.[113]

SARs are one small part of a wider system of watchlisting that draws in the CIA, the NSA, and the FBI. The government's Terrorist Screening Database (TSDB) contained a list of 680,000 people as of 2014.[114] The National Counterterrorism Center defended the breadth of the system, noting that prior to the electronic system, most watchlists were handwritten. As it turns out, even this terrorist watchlist—which contains a range of biometric data—is smaller than the Terrorist Identities Datamart Environment (TIDE) database, which uses a more expansive, laxer set of rules for adding names to the list. In the summer of 2013, it reached one million names. The worldwide individualization of warfare here finds its most sophisticated policing apparatus: not nation-states but individuals—existing anywhere—are the targets of this expansive surveillance.

Indeed, the sophistication of U.S. policing has developed in tandem with advances made in biometrics across the globe. Historically, a crucial leap was made after the information revolution of the late nineteenth century.[115] The photographic identification system pioneered by Alphonse Bertillon in the late 1800s, together with Sir

Edward Henry's method of fingerprinting in the British Raj, became standards for policing in major U.S. cities by the turn of the twentieth century, establishing a criminal-history system. But that's not all. The physical architectures of policing were becoming smarter in the wake of the electronic revolution. Centralized phone systems replaced telegraphs, thereby connecting police stations across the city. "By 1900, American cities had a total of 912 municipal security networks transmitting forty-one million messages annually."[116] The city, in other words, became *wired,* and the arteries of surveillance slowly implanted themselves in the urban flesh.

The mug shot and fingerprinting remain central to policing. The FBI's Integrated Automated Fingerprint Identification System (IAFIS) is the biggest fingerprint database in the world, with over one hundred million criminal and civil records. By 2011 it was processing around 168,000 fingerprint check requests every day.[117] The IAFIS is accompanied by two large civilian databases: the Department of Homeland Security's Automated Biometric Identification System (IDENT), which is used at border points, and the Department of State's Consular Consolidated Database (CCD), which screens photographs of visa applicants. Biometrics continue to advance. The FBI's Next Generation Identification (NGI) biometric database expands upon the FBI's current system. The NGI, which came online in 2014, holds multiple forms of biometric data, including iris scans and palm prints and photographs of facial scars, unusual marks, and tattoos (a service called Scars, Marks, and Tattoos (SMT), which is part of the Interstate Photo System).

Crucially, the NGI can automatically recognize human faces from its database. By 2013 the NGI database held sixteen million images, representing around eight million individuals, and was capable of conducting tens of thousands of daily searches.[118] By 2015 the NGI aims at including fifty-two million photographs. Controversially, the NGI will include images of Americans who have never committed a crime. Currently, if an individual applies for a job in the United States that requires a background check, their fingerprints are sent to and stored in the FBI's civil print database. Photographs were not previously part of the record. This will change with the NGI. If an employer requires a mug shot to accompany an individual's fingerprints, the FBI will store that data. This information will then be fed into a database that merges criminal and civilian biometrics. Records will be assigned an FBI Universal Control Number, and police searches will run against all of the records held by the NGI. Relatedly, the growing use of CCTV systems is creating the infrastructural basis for facial recognition on the fly. With cameras capable of gigapixel technology, the days of public anonymity are coming to a close. Networks of interoperable CCTV cameras will be able to recognize, identify, track, and, increasingly, *predict* the movements of city dwellers as they navigate their daily grind. Topographic distance collapses under the topological wormholes opened up by this algorithmic governance.

While the bodies and cars of police officers in streets remain important vectors of state power, U.S. policing increasingly follows a technologically intensive model based on dominating the city's patterns of life with a mangle of machines. Increasingly, these systems are built to *seize the future*. Phillip K. Dick's short story "The Minority Report" explores a world in which the police are able to accurately predict and stop a crime before it is committed. The conceit throws up all kinds of interesting philosophical questions, such as the paradox of punishing a person for an act they did not—and now cannot—perform. Whereas this piece of science fiction is based on the clairvoyant minds of "precogs," computer algorithms are now able to predict and *secure* the future based on modeling vast sums of data.

Just as IBM computers underlined Igloo White in the late 1960s, its technology also lies behind a form of future policing in the United States in which proprietary algorithms are used to predict and geolocate the emergence of crime hot spots. An important precedent belongs to the New York City Police Department, which first used the CompStat program for a data-driven approach to policing in 1994. Smart policing, advertises IBM, means fighting crime with data.[119] In 2005 the Memphis Police Department (MPD) developed Blue CRUSH, or Crime Reduction Utilizing Statistical History, with the University of Memphis. The pilot scheme used IBM statistical software and GIS programs to predict the geography of potential crime before it happened. The experiment apparently yielded impressive results. After a few months 1,200 arrests were made based on the data. Since 2009, the National Institute of Justice has made millions of dollars available in grants for extending predictive policing across the United States.[120] The LAPD has trialed its own version of future policing using PredPol software. As with Blue CRUSH, computer algorithms are used to predict the probabilities of future crime in space and time. PredPol automatically calculates a hot spot measuring five hundred by five hundred feet, which then guides police officers on their shift. "I'm convinced that predictive policing works," says LAPD captain Sean Malinowski.[121]

Unlike in "The Minority Report," states one of PredPol's developers, "this is about predicting where and when crime is most likely to occur, not who will commit it."[122] But predictive policing isn't limited to geolocation. Beginning in the summer of 2013, the Chicago Police Department (CPD) began trialing a new initiative that targeted individuals whose futures were at risk. The program created a Heat List of around 420 individuals who were predicted to be involved in gun crime sometime in the future. Officially, the program is called Custom Notification, and the strategy is based on police officers notifying an individual that further criminal activity, even for petty offenses, holds grave consequences. Based on numerous variables, particularly an individual's known acquaintances and criminal history, the computer program predicts these at-risk individuals are five hundred times more likely to be involved in violence than are others in the general public. Commander Lewin, who heads CPD's information technology, writes, "This will inform police departments around the country and around the world on how best to utilize predictive policing to solve problems. This is about saving lives."[123] In one instance the precrime list was prophetic: an individual was shot as he walked home in his South Chicago neighborhood.[124] But the slope here is slippery. Although the Custom Notification program may have good intentions, where exactly does it stop? Is it so outrageous to imagine preemptive detention if a computer algorithm calculates an 85 percent chance an individual will commit a shoplifting offense later in the month? The U.S. military and CIA already *kills* by algorithm. Arresting by algorithm is no leap.

In 2009, as part of its growing commitment to predictive policing, the LAPD opened the Real-Time Analysis and Critical Response Division. The 84,000-square-foot regional crime center pulls together facial-recognition software, data mining, and geographical profiling for up-to-date actionable information. Feeding this nerve center is a growing list of surveillance infrastructures. The LAPD center has access to over one thousand CCTV cameras, live feed from social media, and automatic number-plate readers. Since these were first used in 2005, local police have logged the movements of millions of drivers across Southern California in one of the densest license plate–recognition systems in the United States.[125] A police officer can then retroactively produce a map of where a vehicle has been (the LAPD stores this information for five years). One sergeant from the sheriff's department stated, "It's not Big Brother. It's doing exactly what a deputy normally does in his routine duties."[126]

But automated surveillance is not just limited to CCTV. The L.A. County Sheriff's Department has tested an airplane-mounted surveillance kit that records, in real time, the activity of the entire city of Compton. The system uses a civilian aircraft fitted with a cluster of high-definition cameras in a setup not dissimilar to the U.S. military's Gorgon Stare technology. Additionally, the program has the ability to rewind the visual feed.[127] It's not just the future that can be dominated but the past, too.

In a little over a decade, an algorithmic war against time has gripped U.S. policing. The police are now able to mine the past, dominate the present, and predict the future. Action thus becomes preemptive in the rush to prevent individuals from transgressing the law. This reflects broader trends in the U.S. military, which is similarly fixated on controlling time. In the rush toward preemptive manhunting, the Pentagon and the CIA have shifted from a status-based form of targeting aimed at identifiable soldiers to an evidence-based form of targeting. This regime secures—and eliminates—those dangerous patterns of life that circulate across the electromagnetic spectrum: a necroforensics that digitizes life to produce death. The rush to secure time collapses war power and police power inside a single electronic battlefield, and it may be difficult to turn back the clock on these apparatuses.

Electronic control is now a principal means by which control societies are policed. As Louise Amoore, Stephen Marmura, and Mark B. Salter argue, "Castles, walled cities, and extensive border battlements have been replaced by gated communities, expansive border zones, and management by 'remote control.'"[128] The urban apartheid of today is evolving, with physical walls being complemented by electromagnetic borders "constituted as much by dataflows, artificial zones and spaces of enclosure that seep into the city and the neighborhood."[129] There is, therefore, an inherent agency to the materiality of the modern, securitized city. The more-than-human urban landscape is active in the monitoring, recording, and policing of life. Rather than passive matter enclosing active subjects (as with the disciplinary society), the city of the control society is built with active matter enclosing passive subjects. Agency is redistributed among the concrete bones and electric nerves of the metropolis. No longer separate from the city, the police are immanent to its concrete folds, dominating its data spheres and regulating the ecologies of human life.

Atmospheric Security, Part I: Urban Apartheid

Over half the world's population—approximately 3.9 billion people—live in an urban environment. This urbanization of the world's population has contributed in turn to the spread of slums. "Instead of cities of light soaring toward heaven," writes Mike Davis, "much of the twenty-first-century urban world squats in squalor, surrounded by pollution, excrement, and decay."[130] The situation is getting worse, as technological civilization becomes ever more capital intensive and governments pursue neoliberal policies that exacerbate urban inequality. "The result," argues Stephen Graham, "is a kind of social, civil war to control domestic space."[131] After agricultural enclosure masses of newly liberated families were thrown into the booming cities of industrial England. The slum was the most obvious spatial footprint of poverty and the beginning of a form of urban apartheid. As Engels observes in his tour of English slums, the town embodied a ruthless individual survivalism: "The dissolution of mankind into monads, of which each one has a separate essence, and a separate purpose, the world of atoms, is here carried out to its utmost extreme. Hence it comes, too, that the social war, the war of each against all, is here openly declared."[132]

To understand the spatiality of this atomized war of all against all—as well as its consequences—we need to rethink the shifting spatiality of human existence. What does it mean to say human existence is spatial or atmospheric? Run or hide, we can't escape the atmosphere that surrounds us. It is central to who we are, and our destiny is bound to the environments we dwell inside. Martin Heidegger's notion of "being-in-the-world"[133] describes the inescapable condition of human beings immersed in a space of activity, people, and things. In describing this condition, argues Hubert Dreyfus, "Heidegger questions the view that experience is always and most basically a relation between a self-contained subject with mental content (the inner) and an independent object (the outer)."[134] The spatiality of being-in-the-world cannot be translated to the way we typically imagine objects placed inside containers. It is better understood as a space of activity, immersion, and dwelling. "*Space is neither in the subject nor is the world in space,*" insists Heidegger. Instead, the subject—if "correctly understood"—is in fact "spatial."[135]

Building upon Heidegger's philosophy but twisting it into new shapes, Sloterdijk argues the spatiality of human existence is being-in-spheres. A sphere, to recall, is the space of human coexistence, the voluminous atmosphere that surrounds us. This idea that the human subject is always with and possessed by an environment forms a radical critique of modernity, which insists upon the autonomy of the individual subject and denies "all relation between content and container."[136]

Sloterdijk is keen to argue the metaphysical layers that once sheltered the ancient Greek cosmos—the enveloping bubbles that surrounded humans on earth—have now been shattered by modern, world-alienating science. The cosmic sphere is gone, and the great religious cupola that once stretched across the skies—housing God and his angels—has vanished. We are no longer immunized against the frost that seeps in from the universe. Instead, humans must now create all kinds of prosthetic shells to defend against the darkness.

The most perfect expression of such an artificial sphere is the space station floating in the void: a model for the future habitat of mankind. "Nobody can say today if the space station will be the future of the human type. But it represents a model for being in a world condemned to artificiality."[137] These marooned capsules deny nature, deny neighbors, and deny society as such. "The urban landscape is shifting from the modern focus on the common square and the public encounter," write Hardt and Negri, "to the closed spaces of malls, freeways, and gated communities."[138] Moreover, as Allen Feldman argues, this means the biopolitics of immunity moves to the urban interior, "creating new boundary systems that are virtual, mediatized, such as electronic, biometric, and digital surveillance nets."[139]

A profound spheric realignment has thus taken place under the modern age. The isomorphism classically shared among sovereignty, nationality, and territory has been shattered by waves of electronic globalization. As a result, the spaces of immunity have progressively shrunk from the imagined community of a shared humanity right down to the individual. The voluminous rings that once surrounded human being, stretching into the clouds, have slowly evaporated into the night. As Sloterdijk explains, "If classic modernism still banked on immunization through the collectivity—the nation-state, proletarian solidarity, intellectual communities—the construction of postmodern immune systems puts more of an accent on individuals."[140] Across technological civilization, the societies of control are "composed wholly of atoms," as Engels so provocatively describes. The lonely individual, now more than ever, bears responsibility for fortressing its world.

As discussed, the model of the Panopticon has spilled out from its state walls to effect a much more universalized model of control based on fixing and fragmenting human conglomerates across the social field. Technological civilization individuates atomized human beings inside these disjointed immunity bubbles. Sloterdijk labels this distinctive existential morphology "foam." As he explains, "I propose the metaphor of foam, a term which in its own polyvalent nature expresses very well the multi-cellular composition of the big, rather amorphous structures that correspond to the populated landscapes of our era, and especially to the urban conglomerates that are like veritable foams composed of individualistic cells."[141]

The modern city is a polyspheric network of separation that embodies the splintering cacophonies of technological civilization: a foam of differential sizes, shapes, and structures that secures and repels in equal measure. The spatial concept of foam is thus indicative of the uneven fortressing of the city, what Francisco Klauser conceptualizes as "splintering spheres of security"—that is, "an ensemble of spatially anchored, more or less hermetically enclosed, socially exclusive, and atmospherically active spheres of togetherness that are, essentially, composed by co-isolated, individuated subjects."[142]

Rather than living in an all-encompassing dome, with its certain and holistic semiosphere, the citizens of technological civilization float in a restless foam, one that is generative of alienated and disaffected psyches. Although we are isolated inside our bubbles, we are coisolated, mutually lonely. The bubbles press against each other, forming fragile and fleeting ensembles. For Sloterdijk the "foam-cell city" is built on the dismantling of social conglomerates into individualized, fragmented, and complex entities. This foamy togetherness is fleeting, purified, and highly controlled. Using the shopping mall as an example, Klauser argues contemporary togetherness is based on "a purified inside that guarantees each individual customer's right to enjoy the mall's promise of enjoyable shopping whilst at the same time, if wanted, remaining undisturbed and left alone."[143]

Cities split between insides and outsides constitute what Lieven de Cauter calls the "capsular civilization."[144] As he writes, "The First world is no longer a homogenous empire covering a relatively homogenous territory, but an archipelago of fortresses and strongholds. . . . The generic city is obsessed by closure, safety and control. One can appropriately term this the *cellular city,* and even the *capsular civilization.*"[145] Enclosure, in other words, is constituted today by an archipelago of fortified capsules. As de Cauter continues, "*Capsule architecture* is the architecture of the generic city. The capsule is a device that creates an artificial *ambiente,* which minimizes communication with the outside by forming its own time-space milieu, an enclosed (artificial) environment."[146] Urban space is thus increasingly polarized between the slum, which houses a planetary surplus population, and the capsular architecture that seeks to isolate a shrinking bourgeoisie. These are two of the spectrums for living in technological civilization.

Capsules include forms of transport. Don Mitchell argues a "S.U.V. model of citizenship" has taken root in the urban United States, realizing what Marx calls the end of civic space and the ascent of "purely atomic social relations." We are free to collide with other individuals, Mitchell explains, "But *we do not want to collide with one another*; we want to move freely through public space, encased in an impregnable bubble of property (made real through law), and watched over by a network of surveillance cameras, their operators, and the state. We want—and expect—to feel safe at all times."[147] Public space becomes a honeycomb of fast-traveling capsules, shielded from the unbearable aura of our neighbors.

As Graham explains, "The city outside is rendered as a brutal, Hobbesian space of threat and fear, while the cocoon within is a safe, civilized, portable refuge."[148] Security and surveillance apparatuses have come to fortify islands of wealth from the oceans of poverty that surround them. The lives of the poor and the rich are being reorganized such that the two masses of humanity never have to interact with each other. The superrich plug into artificial, securitized cocoons, transferring bodies, minds, and capital into a topology of capsules protected against the outside. "As ever more capsular and lavish domestic spaces, with their mythic allure of certainty, homogeneity, order and control, are constructed," argues Graham, "they are being surrounded by configurations of attempted withdrawal from the risky, racialized, and often poverty-stricken open city."[149]

Enclaves, enclosures, capsules, and insides, all of these spatial concepts are diagrams for taking seriously the patchwork of security atmospheres that subdivide the urban landscape.

Furthermore, Sloterdijk argues "verticalization" is a key axis of contemporary urban immunity. As he writes, "The redefinition of urban space emerges on stilts: Above the desperately exposed cityscapes of the status quo, tall pillar-systems radically elevate artificial and new spatial articulations, in which the urbanites of the future can exist among each other."[150] Modern elites attempt to extricate themselves from the messiness of the earth below. "Might the global proliferation of iconic as well as more prosaic high-rise residential, corporate and hotel skyscrapers," ask Stephen Graham and Lucy Hewitt, "contribute in many cities to the emergence of a myriad of vertically stratified, gated 'communities' which residualize the surface city as powerfully as exurban gated communities residualized traditional public street systems?"[151] Their point is that vertical structures like skyscrapers act as insulated capsules, cocoons that secure socioeconomic elites from the insecure surplus populations below. The safest—that is to say, most enclosed—planet we've ever lived on is nonetheless deeply disaffected.

The generic, mass-produced capsules—from cookie-cutter homes to SUVs—produce a widespread conformity, individuating feeble herds inside privatized spheres while terrifying them about the hell that exists beyond the walls. What de Cauter calls the "hypertrophy of the private sphere" is symptomatic of a retreat from what Hannah Arendt calls the "polis" and thus represents a dangerous withdrawal from politics. As de Cauter pithily states, "Capsular architecture is ostrich politics."[152] Capsularization is thus productive of what Arendt calls the "mass man." Anxiety, fear, and depression have become entrenched phenomenological states in technological civilization. Accordingly, there is an important psychospherical aspect to foamy, capsular existence. As masses of humans are immunized inside their artificial husks, an extraordinary psychic individuation is executed across the species.

As both Heidegger and Sloterdijk outlined, the essential idea contained within the spatial metaphor of the sphere, the capsule, or the atmosphere is the fundamental disruption of the division between individual and environment. Both object and subject stand outside themselves. Affective atmospheres, argues Ben Anderson, "occur *before* and *alongside* the formation of subjectivity, *across* human and nonhuman materialities, and *in-between* subject/object distinctions."[153] Atmospheres, in other words, crystallize the human and nonhuman forces that swirl in their midst. "On this account atmospheres are spatially discharged affective qualities that are autonomous from the bodies that they emerge from, enable and perish with."[154] In the capsules of technological civilization, an anxious, paranoid atmosphere hangs in the volumes of our coexistence. Moreover, since the Predator Empire has become a global force dominating a battlespace immanent to the social field, "trauma has moved away from the battlefield and into every walk of life."[155]

Atmospheric Security, Part II: Military Urbanism

The geopolitical implications of a world in which an insulated bourgeoisie militarizes a planet of jobless cities are profound. For most of human history, empires had to pacify rural counterinsurgencies. This has now changed, with the "the poor peripheries of developing cities" becoming what Davis calls "the permanent battlefields of the twenty-first century."[156] Accordingly, more explicit forms of urban military intervention have begun to take hold. The city condenses the spatial contradictions of technological civilization to the street level: a frenzied honeycomb of insides and outsides, or valued and surplus populations. This urban social war has in turn become increasingly weaponized, particularly during the war on terror. As Graham argues, "Everywhere, the urban boundaries between the 'insides' and the 'outsides' of our planet's dominant economic order present sites of palpable militarization."[157] Under this understanding, the nation-state is no longer a singular container of sovereign immunity and protection. National borders are reemerging in the policing and securing of urban enclaves. As de Cauter argues, "Whereas the disciplinary society was based on internalization, the control society functions externally, through militarization of urban space."[158]

U.S. military tactics are now geared toward finding, fixing, and finishing suspicious patterns of life in the city, what Graham calls the "new military urbanism."[159] Military Operations (MO), or what the U.S. military previously called Military Operations on Urbanized Terrain (MOUT), embody the tactics used to subdue the insurgent city. The U.S. Army's Field Manual 90-10 was first released in 1979 and was an early attempt to theorize combat on the "urbanized battlefield."[160] By 1996 Major Ralph Peters argued, "The U.S. military must stop preparing for its dream war and get down to the reality of the fractured and ugly world in which we live—a world that lives in cities."[161] Dystopian visions of looming urban battlespaces—partly a result of the 1993 Black Hawk Down debacle in Mogadishu—were reanimated by the urban counterinsurgency pursued in the war on terror. Militarized techniques of targeting have come to colonize the urban landscape. Airports, train stations, hotels, and city centers are part of a globally resonating battlespace. Such a form of security is based on profiling, anticipating, and separating "risky" lives wherever they may move.

This includes Western cities, which have increasingly adopted the counterinsurgency tactics of "explicitly colonial models of pacification, militarization and control, honed on the streets of the global South."[162] "Enemy" is now a much more fluid category, whose violation can be as menial as the transgression of the neoliberal order. There thus exists a generalized social war inscribed into the material backdrop of the city: a war against crime, a war against drugs, a war against homelessness, a war against terrorism, and of course, a war against poverty. Feldman calls this a type of "securocratic war,"[163] or a war for public safety: "These wars are not exclusively focused on territorial conquest, or an easily locatable or identifiable enemy with its own respective goals of territorial appropriation. Rather, they are focused on countering imputed territorial contamination and transgression—'terrorist,' demographic and biological infiltration. These campaigns are not structured by time-limited political goals but are temporally open-ended."[164]

In turn, Graham argues welfare states are "being re-engineered as risk-management systems, geared not towards the social welfare of communities but toward controlling the location, behaviour and future of seemingly risky 'anti-citizens.'"[165] This not only divides the planet into areas of safety and risk but calcifies a splintered, capsulized city, or what Feldman calls a "dual city," which is "enforced and reproduced by technologies of spatial control."[166] In short, urban space has become a crucial domain for leveraging and executing military power. The enclosure of the city—by soldier, CCTV, or drone—relies upon a system of militarized enclaves, securitized corridors, and automated control systems.

Although the term *atmospherics* originates in the U.S. military to describe the feel and minutiae of a hostile, occupied environment, the term is equally applicable to homeland security, which is immersing itself in the atmosphere and becoming more atmospheric. "Thinking through atmospheres," argues Peter Adey, "helps us to extrude networked security into immersive enveloping shapes in which the subjects in and of security might be caught. Thus, understanding a security dispositif through atmosphere is to attend to the multiple, expressive, and enveloping spatialities it produces and seeks to capture."[167] The city has become a crucial target for atmospheric forms of state power. In the great metropolises of technological civilization, the minutiae of daily mobilities are condensed in a totalizing form of surveillance. If atmosphere suggests the enveloping shapes in which we are caught, then consider the sheer ubiquity of urban technologies that gather human patterns of life. The citizen becomes a virtual citizen moving across an invisible and remote cloud, a "holographic composite or profile of gestures, data, and algorithms."[168]

In addition to infecting the city's atmosphere with a technological DNA, there is another sense in which state power is increasingly atmospheric. If, as Weizman suggests, geopolitics is a "flat discourse" bound to the limits of ancient maps, then a politics of verticality "requires an Escher-like representation of space, a territorial hologram in which political acts of manipulation and multiplication of the territory transform a two-dimensional surface into a three-dimensional volume."[169] Sovereign power, under this understanding, must control the atmospheres that surround the spaces below. Building upon this vertical understanding, Stuart Elden argues territory is a *volume* rather than a flat plane. As he reflects, "What happens if we take the vertical as a key question, taking the additional dimension into account if security has to contend with *volume*? What would it mean to 'secure the volume'? How does thinking about volume—height and depth instead of surfaces, three dimensions instead of areas—change how we think about the politics of space?"[170]

A modern example of volumetric security has been the continued enclosure of airspace in the West Bank and Gaza with military satellites, jets, drones, and blimps. This has allowed the Israeli military to tap into material and electronic circulations of Palestinian life. "In a 'vacuum-cleaner' approach to intelligence gathering, sensors aboard unmanned air vehicles (UAVs), aerial reconnaissance jets, early warning Hawkeye planes, and even an Earth-Observation Image Satellite, snatch most signals out of the air. Every floor in every house, every car, every telephone call or radio transmission, even the smallest event that occurs on the terrain, can thus be monitored, policed or destroyed from the air."[171]

No longer is sovereignty limited to the cartographic surfaces of the planet; it has now ascended into the atmosphere, seeking to enclose the totality of human communication. This kind of atmospheric security can enclose entire cities below, with hundreds and, at times, thousands of people existing with the threat of death hanging above their heads. Sentenced to life inside a paranoid dome, there is little respite for the groundlings enveloped inside these imperial spheres. Indeed, the atmosphere is now an explicit medium for security and violence, or what Sloterdijk calls "atmoterrorism," a type of terror that uses the air as both a medium of violence (as with drones traveling in the skies) and a weapon (as with chlorinated gasses). As Sloterdijk argues, "The art of killing with the environment is one of the big ideas of modern civilization."[172]

Vietnam was the site of a profound experiment in geographic warfare. The environment was explicitly targeted by the U.S. military: Agent Orange, bulldozers, napalm, and aerial bombing re-engineered the territory to fit the abstract blueprints of technowar. Although not as destructive, the use of tear gas during the war holds an important connection with modern U.S. policing. Tear gas is made from various chemical compounds, and most types are designed to irritate the mucous membranes in the eyes, nose, mouth, and lungs, leading to respiratory difficulties and temporary blindness. In addition to creating an intense burning pain, these so-called riot-control agents (RCAs) can be psychologically terrifying. As the gas indiscriminately engulfs a crowd, an involuntary panic can grip those trapped in its plumes. As the U.S. Army's manual on civil disturbances warns, "The use of riot control agents (RCAs) by authorities adds to the panic and confusion."[173] Tear gas has become an important weapon for securing the urban milieu.

After being used in Vietnam in the 1960s, tear gas was used by the Royal Ulster Constabulary in Northern Ireland during the Troubles, with the first recorded usage by British police in August 1969.[174] Over the next decade tear gas was used across the planet, from South Korea to Palestine. More powerful chemicals and new methods of delivery were developed during the 1990s, and while the 1993 Chemical Weapons Convention reiterated an international ban on tear gas during warfare, it made an exception for the use of tear gas in riot-control situations by police and law enforcement. "In the 100 years since it was first developed," argues Anna Feigenbaum, "tear gas, advertised as a harmless substance, has often proven fatal, asphyxiating children and adults, causing miscarriages, and injuring many."[175]

By flooding the urban volume with a dense fog of chemicals, the state is able to quickly seize the territory that protestors occupy. The Arab Spring and the regional revolts that followed it witnessed the large-scale gassing of protestors. In Bahrain, for example, the atmoterrorist state used excessive volumes of tear gas in 2011 and 2012, leading to the suspected death of thirty-four Bahrainis.[176] Marijn Nieuwenhuis concludes, "The ongoing struggle over Gezi Park and other spaces around Turkey [another site of widespread teargassing] (and beyond) are no longer primarily being fought on the ground. They are instead increasingly taking place in the air."[177] This atmoterrorism is justified frequently by the police in the name of maintaining liberal democratic society or some form of civil peace. Of course, the use of tear gas deliberately blurs the apparent division between war and peace. Since tear gas is an atmospheric weapon launched *against* and *through* the atmosphere, the urban instantly becomes a battlefield, since it divides those who can occupy the space (usually police with gas masks) from those who cannot.

The enveloping, voluminous quality of gas transforms the urban ecology into a hostile milieu. "In this war, the atmosphere itself becomes the war theater. More: the air becomes a weapon and a battlefield of a peculiar kind," Sloterdijk argues.[178] Unlike the strictly Foucaultian disciplinary power that targets bodies or the biopower that targets populations, atmotechnics targets both by polluting the atmosphere and infecting the shifting, life-sustaining spheres of coexistence. Sovereign law is therefore enforced directly upon bare, animal life—in the lungs of the unwilling subject. Biological tissue becomes a target and a participant in the modern science of atmospheric warfare. "Living together," writes Philippe Theophanidis, "has become the environment in which the political management of life takes place as the possibility [of] life's own annihilation."[179] In other words, the living environment is turned against those who dwell inside it. Our very being-together is conditioned by violent forms of state-directed pacification. Under an intensifying form of military urbanism, territory is secured by totalizing forms of atmospheric power that target the city's volumes.

Atmospheric Security, Part III: The Dronification of the Atmosphere

The final form of spheric securitization is the use of drones to pervade, police, and petrify the atmosphere. In the case of those human beings watched by Predators in Pakistan, Yemen, or Mali, a verticalized war machine already translates their lifeworld into electronic wraiths passing between screens. Drones tear open the sensual world of a population only to reconfigure it within the imperial geometries of an enclosed security atmosphere. Drones, like any machine, process worldly information. State power is invested precisely in the moment between input and output—the passage through which reality is operated on, engineered, or transduced.

The drone is a crucial machine for realizing this spheric hacking, probing beneath rooftops with infrared cameras, recording the minutiae of daily life, and occasionally raining Hellfire. The drone crystallizes affective relations around the technological order it authorizes. By occupying the sky, a form of terror saturates the ground below like a miserable fog. "People are afraid of dying," said one shopkeeper in Pakistan, "Children, women, they are all psychologically affected. They look at the sky to see if there are drones. [The drones] make such a noise that everyone is scared."[180] Human lives become *fossilized*—that is to say, daily existence is petrified as a form of living death. Aerial policing is as much a psychological power as it is a geopolitical force, and drones invade minds as much as they do atmospheres: individuating a paranoid and enfeebled subject. Thinking about drones as spheric weapons therefore expands the existential dimension of the Predator Empire.

As drone technology has fallen in cost, a variety of hobbyists, lobbyists, and vested interests has pressured the government to open American skies to drones. Of course, flying drones in civilian airspace creates a number of safety issues. Between 2013 and 2014, there were fifteen cases of drones flying dangerously close to passenger aircraft and hundreds of other unsafe incidents.[181] Currently, drones are granted access to U.S. airspace through the issuance of a Certificate of Waiver or Authorization (COA). At some point in the near future, this ad hoc approach will change to an integrated system of rules and regulations, as pursuant to the congressional Federal Aviation Administration (FAA) Modernization and Reform Act of 2012. It will be difficult, however—impossible even—to track each and every drone, despite the FAA's recent requirement that small drones be registered. The FAA is therefore keen to develop sense-and-avoid algorithms to enable future drones to autonomously reroute their flight paths. According to the FAA's Destination 2025 roadmap, this means the Next Generation Air Transportation System will eventually be required. Imagine a kind of futuristic Skynet (the ubiquitous system of artificial intelligence from the *Terminator* films) that can augment and unburden human operators.

So far, hunter–killer drones have operated outside the United States and will continue to do so for some time. Nonetheless, a big trend in recent years has been the adoption of drone technology within U.S. law enforcement. Such a sociotechnological momentum raises serious questions. The FBI has flown drones for surveillance operations on at least four separate occasions since 2010.[182] Between 2004 and 2013, the bureau spent around $3 million developing drone technology.[183] Funding for drones and other robotics has often been channeled through the Department of Homeland Security's Urban Areas Security Initiative and the 1033 Program. These grants have enabled police departments across the United States to launch their own drone programs. A big user of Predator drones has been U.S. Customs and Border Protection (CBP), which has operated around ten Predators along the U.S. borders with Mexico and Canada since 2004. In 2012 its fleet of drones flew for approximately 5,102 hours.[184] Additionally, CBP loaned its Predators to law enforcement agencies seven hundred times over a three-year period beginning in 2010.[185] In North Dakota in 2011, CBP Predator drones were involved in a local police force's arrest of a U.S. citizen, the first incident of its kind.

This arrest symbolizes the kind of mission creep inherent to the use of drones, from the narrowly defined assignment of securing the U.S.–Mexico border to a more generalized law enforcement from the skies, one that includes surveilling the everyday lives of Americans. Already, incidental data that CBP Predators intercept during an investigation can be stored indefinitely, "even if it includes footage of property, vehicles and people unassociated with the investigation."[186] There have even been DHS proposals to modify Predator drones so they can track cell phones.[187] Once local cops have their own drone fleet, argues Naomi Wolf, "the meshing of military, domestic law enforcement, and commercial interests is absolute. You don't need a messy, distressing declaration of martial law."[188] The drone society simply mirrors the colonial battlefields abroad. As such, it is likely that police drones will one day be fitted with nonlethal weaponry to subdue civil unrest. At least in the beginning, they'll deliver tear gas instead of missiles. Such bleak—and admittedly dystopic visions—are not simply fantasy. As U.S. colonel Carlile reminds us, "The difference between science fiction and science is timing."[189] And of course, that's not to rule out weaponized drones. As Greenwald argues, "American surveillance drones went from Yemen, Pakistan and Somalia into American cities, and it's impossible to imagine that they won't be followed by weaponized ones."[190]

Of course, it's not just militaries and law enforcement that will use drones. There has been an explosion in the number of drones used to ferry narcotics across national borders, state lines, and even prison walls. Unfortunately, it's just a matter of time before lone wolves use drones to strike high-value targets, politicians, and highly populated urban areas in the U.S. homeland. In response American cities will be further militarized and surveilled as no-drone zones emerge around corporate and government buildings. There is already a market for electronic antidrone jammers (such as the Droneshield early warning system). The White House is developing this kind of technology after a drone crashed on its lawns in 2015. Relatedly, weapons contractor Raytheon launched a blimp system able to create a vast surveillance radar net over a 340-mile stretch of America's northeast skies. Military blimps have long been an established technology for surveillance in Afghanistan. The aerostat system in Maryland is called the Joint Land Attack Cruise Missile Defense Elevated Netted Sensor System and costs a staggering $2.7 billion.[191] From a ten-thousand-foot vantage point, the blimps—flown from the army's Aberdeen Proving Ground—will be able to detect enemy aircraft, including drones, before they reach the U.S. capitol. The ACLU warned, "That's the kind of massive persistent surveillance we've always been concerned about with drones."[192] The potential uses of drones—whether good, bad, or ugly—are as limitless as the human imagination. Tens of thousands of drones will spill into the skies within decades, delivering pizza and teargassing protestors. Indeed, the corporate sector is driving the drone society as much as the military is.

The end goal, as far away as it may be, is to install a system of ubiquitous air policing across major U.S. cities. The Predator Empire therefore becomes a permanent police presence in the skies. As Neocleous argues, "One might want to say that the City has become a war zone subject to absolute police power. But then one might also want to say that the City has become a police zone subject to absolute war power. One would be making the same point either way."[193] The danger, then—and it's worth repeating Chalmers Johnson on this point—is that "even an empire cannot control the long-term effects of its policies. That is the essence of blowback."[194] The robots in distant battlefields always return home.

Swarms of police drones are coming. The electric plasma of our atmosphere will soon swim with robots. A militaristic dream of continuous, automated, robotized counterinsurgency warfare is upon us, with "myriads of robotic devices spread generously through the 'urban battlespace' which use computer code linked to vast databases to automatically define and even destroy 'targets.'"[195] Infrared technologies are already turning our homes into glass houses, and the NSA intercepts much of our telecommunications. Drones in the skies—constantly monitoring, recording, tracking, policing, subduing—are the endgame of the robot imperium. In 2011 the Pentagon began research into drones capable of sophisticated tagging, tracking, and locating (TTL)—drones that never forget a face. "If this works out, we'll have the ability to track people persistently across wide areas. A guy can go under a bridge or inside a house. But when he comes out, we'll know it was the same guy that went in."[196] The Long Range, Non-cooperative, Biometric Tagging, Tracking and Location system uses algorithms to convert two-dimensional images taken by a drone into a three-dimensional model of a face. Another project funded by the army seeks to determine "adversarial intent" based on monitoring human behavior.[197]

For years drones have promised radical changes to the conduct of conflict. A big push in the future battlespace is for the integration of drones across each branch of the military. A recent U.S. military roadmap paints a vivid picture of an automated battlespace that unites birdlike drones perched on electric power lines with insect-like drones crawling across the ground.[198] This would be a common robotic ecosystem that could operate seamlessly across land, sea, and outer space. Currently, most unmanned systems are remotely piloted, and a human is in the loop. But that will change as military drones shift from being remotely piloted technologies to autonomous systems. The pilot will play more of a supervisory role, pointing and clicking GPS directions to a swarm of intelligent drones.

Back in 1972, after the revolution in drone warfare in Vietnam, one DARPA engineer foresaw this exact future. He predicted that "enemy warplanes will be met by hundreds of RPVs equipped with rockets, missiles, and laser ray guns, swarming around their larger but slower adversaries like bees attacking a bird."[199]

This kind of hive intelligence reaches its apex with the SWARM capability (smart warfighting array of reconfigurable modules). These predictions are not new, of course. "Swarming," John Arquilla and David Ronfeldt of RAND, wrote back in 2000, "is seemingly amorphous, but it is a deliberately structured, coordinated, strategic way to strike from all directions, by means of a sustainable pulsing of force and/or fire, close-in as well as from stand-off positions. It will work best—perhaps it will only work—if it is designed mainly around the deployment of myriad, small, dispersed, networked maneuver units."[200] Guerilla warfare is one of the most persistent forms of swarming. Recently, however, following a succession of information revolutions, the U.S. military has been developing war technology that can swarm. What Arquilla and Ronfeldt call BattleSwarm is a military doctrine based on autonomous, small-scale units engaging in amorphous but coordinated strikes that can rapidly blanket a nonlinear battlespace from multiple directions. In other words, swarming is a form of atmospheric attack. The entire volume is saturated with and secured by fast-moving drones. Indeed, the U.S. Army has long envisaged these smaller drones, or Nanos, interacting with each other in marauding clouds. Onboard software is being developed to allow Nano drones to act autonomously and cooperatively in emergent, nonlinear patterns across a dense urban battlespace.

As one army roadmap predicts, "By 2025, Nanos will collaborate with one another to create swarms of Nanos that can cover large outdoor and indoor areas. The swarms will have a level of autonomy and self-awareness that will allow them to shift formations in order to maximize coverage and cover down on dead spots. Nanos will possess the ability to fly, crawl, adjust their positions, and navigate increasingly confined spaces."[201] These Nanos include drones that mimic insects, drones that are able to perch and hover in the air like birds, and cyborg drones that meld flesh and metal together. By going smaller, drones are able to saturate the atmosphere and overwhelm a target. As DARPA argues, current weapon systems are "too expensive," and by the time they are operational many are "obsolete." It therefore imagines a "systems of systems" approach in which air superiority is won through a modular, distributed system of interchangeable manned and unmanned aircraft.[202] Here, mass is mobilized through low-cost, even disposable weapons and surveillance systems.

If the pathway from fiction to fact is indeed only a question of timing, then *swarm policing* is perhaps only decades away from spawning on the streets of the United States. This would be a form of total policing that blended precrime reporting, algorithmic surveillance, militarized infrastructures, and atmospheric power: a society of drones policed by drones. Neighborhoods would be turned into glass orbs rendered transparent to a totalizing form of airpower that vacuumed patterns of life, stored terabytes of electronic data, and recorded conversations between friends—and all of this would occur in real time. Whatever the exact size and shape of the Predator Empire in ten or even twenty years, drones will be central to the exercise of state power. "By 2020," to recall McCoy, "the United States will deploy a triple-canopy aerospace shield, advanced cyberwarfare, and digital surveillance to envelop the earth in a robotic grid capable of blinding entire armies on the battlefield or atomizing a single insurgent in field or favela."[203] This "everywhere war"[204] will keep on grinding in the gears of technological civilization. And we are only at the dawn of the integration of robotics, nanotechnology, biotechnology, and the age-old art of killing.

As the control societies of technological civilization continually divest in social security, they never cease to invest in more ingenious forms of national security. Every drone that becomes smarter and deadlier not only drags us closer toward a robotic hell but also embodies a damning indictment of a society that is fundamentally alienated. As drones begin to spill into skies across technological civilization, we may yet appreciate the existential changes of life inside the Predator Empire. But time is running out. The streets are beginning to swarm with robots, and there are fewer places left to hide in the civilizatory enclosures.

Imagine, then, a blueprint for a city of the future: a city secured by a robotic police force hell-bent on erasing the possibility of disorder. Imagine the *dronepolis,* the city of the drone. The dronepolis is the latest in a long succession of urban forms that have pacified and policed the planet's surplus populations. It advances the logic of the machine-readable smart city to its dystopic conclusion: a technologically infused urban apartheid. The lives of the valued and the surplus would be proximate topographically but separated by atmospheric technics. The dronepolis materializes the logic of a robotic form of "exclusionary surveillance"[205] that secures segregation. In the atmospheres of this desperate city, hypermobile police drones surround and enter the homes and workplaces of suspects, in a manhunt in which the human is transformed into an abstract pattern of life: a digital simulacrum chased across the datasets of the control society. The dronepolis does not represent a decisive break from the past but is a new materialization of an already existing social war between a fortified bourgeoisie and a planetary surplus population—one in which the lines between the war on terror, the war on drugs, and the war on poverty dissolve into an amorphous battlespace. And this occurs, increasingly, *everywhere,* as the technogeographies for a dronified city march across the planet, fusing colony and metropole in a world system of enclosure. "Oligarchic capitalist class privilege and power," writes Harvey, "are taking the world in a similar direction almost everywhere. Political power backed by intensifying surveillance, policing and militarised violence is being used to attack the well-being of whole populations deemed expendable and disposable."[206] Describing the ascendance of dronified policing, Neocleous writes, "This is nothing less than a permanent police presence of the reproduction of order—air power as the *everywhere police*—in which the exercise of violence is an ever-present possibility."[207] And this ever-present possibility of violence generates a landscape of psychological terror, further wounding the mental health of a disaffected society. The dronepolis is the city of a technological civilization reaching its most capital-intensive and draconian stage: the city living beneath the long shadow cast by the Predator Empire, the city of a surplus and hypersecured humanity.

Conclusion

The War of All against All

The Mechanical Monster

The Predator Empire is a concept used to describe the contemporary and future U.S. national security state, an arrangement of military power, state violence, and unprecedented surveillance technology. Emerging after President Truman signed the National Security Act of 1947—which created the CIA, the U.S. Air Force, and the Department of Defense and formalized Cold War strategy—the national security state has grown exponentially. "From the Pentagon to the Department of Homeland Security to the labyrinthine world of intelligence, the rise to power of the national security state has been a spectacle of our time," writes Tom Engelhardt.[1] The Predator Empire is a Leviathan, a monster built with the steel and flesh of millions of different actors, spread across land, sea, space, and cyberspace and able to execute a dangerous individual from thousands of miles away. Its name is symbolic of the materialization of a mode of state power (policing), a military strategy (predation), an archetypal technology of remote surveillance (the Predator drone), and a geographical scale (the planetary). All of these—policing, predation, the Predator, and the planet—converge around the belief the U.S. military is the premier guardian of civilization, a theme that has in some way persisted since the Cold War.

Beginning a week after the September 11, 2001, terrorist attacks, President Bush argued the war on terror was not just an American fight: "This is the world's fight. This is civilization's fight."[2] In a 2006 address to the nation, he repeated the same motif, stating, "This struggle has been called a clash of civilizations. In truth, it is a struggle for civilization. We are fighting to maintain the way of life enjoyed by free nations."[3] For Mark Neocleous this means the war on terror is "civilization's return, writ large."[4] "The world is a battlespace" expresses the Predator Empire's ambition to pacify any location on the planet perceived as hostile to the homeland. It is, as Michael Hardt and Antonio Negri write, "a regime that effectively encompasses the spatial totality, or really that rules over the entire 'civilized' world. No territorial boundaries limit its reign."[5] The Predator Empire immunizes not simply the U.S. homeland, then, but a more generalized system of prosthetic, enclosed living this book labels "technological civilization." Indeed, in so many ways this form of American exceptionalism—or what Andrew Bacevich calls the Washington Rules—has been a recurring feature of U.S. geopolitics ever since the birth of the republic (which was itself built upon the mass enclosure of an entire continent).[6]

As the U.S. military has continued to withdraw troops from locations such as Iraq and Afghanistan (replacing them with planes and drones), the idea of policing and protecting civilization persists. In 2015, for example, Senator Richard Burr, chair of the powerful Senate Intelligence Committee, said, "When you look around the world, whether it's in Yemen, whether it's in Syria, whether it's in Iraq, whether it's in Afghanistan or in North Africa with Boko Haram, we've got terrorist elements that are carrying out terrorist acts and if you put that collection together, what you've got is a war on Western civilization. It really doesn't matter which terrorist group we insert into the blank."[7] The war on terror, with this understanding, continues to be a project of securing the insecurities that exist outside the world interior of civilization. It really doesn't matter which terrorist group is inserted into the blank: the empty space Senator Burr conjures is the wild, exceptional void of civilization through which the gears of war and law revolve. U.S. drone warfare—a spreading system of aerial policing—is the latest national security project to pacify and eliminate threats to civilization.

While recognizing these geographical imaginations and the work they do, this book attempts to move beyond a simple understanding that pits civilization against external enemies. Chapters 4 and 5 in particular were engineered to reveal the endemic social war that plagues the world interior of a robotizing technological civilization. Moving between a history of enclosure, surveillance, and modern U.S. policing, I have argued sovereignty is founded upon a fundamental fracture in society that breeds a deep sense of dissatisfaction and alienation among individuals. This complicates the idea that the war on terror protects a perfect communion: it does not. It defends a deeply scarred and unequal constellation of human beings enclosed in conditions deleterious to their physical and mental health. The homeland is, in other words, a deeply scarred place, and the noose of the Predator Empire is tightening around the most intimate spaces of everyday life.

This stranglehold is exactly what Roberto Esposito identifies when he describes the war on terror as an autoimmune disease. "Just as in the most serious autoimmune illnesses, so too in the planetary conflict presently under way: it is excessive defense that ruinously turns on the same body that continues to activate and strengthen it. The result is an absolute identification of opposites: between peace and war, defense and attack, and life and death, they consume themselves without any kind of differential remainder."[8] Rather than preserving life and liberty, the excessive defense of civilization paves a direct pathway to a vicious form of totalitarianism.

The point here is that although al-Shabab in Somalia and al-Qaeda in the Arabian Peninsula may not pose an existential threat to the U.S. homeland—much less to technological civilization—the actual danger posed by such groups matters little to the civilizing offensive. "A classic mistake of empire managers," writes Chalmers Johnson, "is to come to believe that there is nowhere within their domain—in our case, nowhere on earth—in which their presence is not crucial. Sooner or later, it becomes psychologically impossible not to insist on involvement everywhere, which is, of course, the definition of imperial overextension."[9] In other words, once the material apparatuses of a totalizing form of sovereignty are installed, the size of the threat needed to trigger some form of intervention shrinks exponentially. There's nothing rational about the Predator Empire—it trades in equal parts absurdity and tragedy. We see this same logic of imperial overextension in the militarization of U.S. policing. Small-scale consensual crimes such as narcotic use in private homes are sometimes greeted with theatrical no-knock SWAT raids, and constitutionally protected protests are policed with robocops armed with assault rifles fresh off the streets of Kabul. The autoimmune disease eats at the living flesh of society.

As the Leviathans of the Predator Empire multiply, however, the form of totalitarianism that emerges will be different from others in the long history of human enclosure. For Engelhardt "the twentieth century was the century of 'totalitarianisms.' We don't yet have a name, a term, for the surveillance structures Washington is building in this century, but there can be no question that, whatever the present constraints on the system, 'total' has something to do with it and that we are being ushered into a new world."[10] To my mind, the totalitarianism we face in the twenty-first century does have a name—the Predator Empire—but it is one devoid of the visceral ideologies and mass slaughter of humans of the past century. The consequences of passing from a disciplinary society to a control society, as well as from a labor-intensive form of state violence to a machine-intensive form of state violence, are important to consider. The Predator Empire marks the culmination of a system of dominance augmented—and increasingly controlled—by machines, computers, drones, robots, and the electromagnetic networks that connect them. The Predator Empire embodies a new form of American empire that has transitioned away from the counterinsurgencies of Vietnam, Afghanistan, and Iraq.

In the introduction I argue Thomas Hobbes's famous depiction of sovereignty—a Leviathan whose body is made of the bodies of citizens and whose head is the sovereign—needs to be thought of in a different, decentered light. Rather than locating the "soul of the Leviathan," Michel Foucault advises we should "study the multiple peripheral bodies, the bodies that are constituted as subjects by power-effects."[11] But these bodies and power-effects must also be understood as nonhuman. Under this new understanding, the Leviathan transforms from a fleshy monster into a cyborgian colossus: a macroactor composed of people—yes—but also of nonhuman elements, technologies, and objects. Sovereignty has always relied upon the nonhuman to stabilize and mediate the daily intercourse of human beings. As Bruno Latour and Michael Callon argue, "The Leviathan is monstrous too because Hobbes built it using only contracts and the bodies of ideal, supposedly naked, men. But since the actors triumph by associating with themselves other elements than the bodies of men, the result is terrifying . . . like the mechanism of a machine."[12] The Leviathan is therefore an important methodological, theoretical, and geopolitical model for understanding the inner workings of the Predator Empire.

An altogether different problem is posed by this revised understanding of the Leviathan, one that neither Hobbes, Foucault, nor Latour necessarily saw coming. What happens when the balance between the human and the nonhuman tips to such an extent the Leviathan becomes a robot? What happens when humans leave—one by one—the loop of sovereignty altogether, becoming a mass of abandoned souls cast outside a system that no longer serves them? What happens when the Leviathan becomes, in other words, a war machine?

We do not have to search far for these answers. Is it not precisely the case that the modern U.S. national security state—the Predator Empire—gives us more answers to these questions than we dare recognize? Do we not see the ascent of control over compassion, security over support, capital over care, and war over welfare? The human condition is enclosed like never before as the spheres in which we are born, become, and die are overrun with security technologies. Indeed, they are rapidly becoming dronespheres. Hannah Arendt is a guiding light in such times and the first to sound the alarm about the coming rule by Nobody in her critique of bureaucracy. But the Predator Empire, with its technical methods of control, poses difficulties even she could not have foreseen in the mushroom clouds of the Cold War.

Arendt is keen in distinguishing between power and violence, two concepts so often seen as synonymous. Power, for Arendt, is an authentic source of collective action based on the mutual agreement of consenting individuals. Violence does not, however, require consent: it is an entirely instrumental force, a tactic or a means that cannot hope to create a better society but only destroy a preexisting one. Although violence may be directed between and against the bodies of humans, it can, unlike power, function without them: in the material infrastructures of the city, in the abstract systems of government bureaucracies, and, of course, in the barrel of a gun. Because violence relies on instruments, warfare has been driven by a series of revolutions in the conduct of violence, from bows to bombs to drones. Beyond Arendt's formulations, we can draw a distinction between power as a human condition and violence as a nonhuman condition.

Since violence is a means with no clear end, it possesses an unpredictable metaphysical quality. Consequently, as Arendt argues, "the means used to achieve political goals are more often than not of greater relevance to the future world than the intended goals."[13] Tools consistently escape human controls and proliferate in unexpected ways, animated by futures that elude the goals for which they were designed. Here, we can picture drone warfare in its fullest light: as a harbinger of existential blowback, the scope of which we may be able to imagine—dronified state violence, dronified criminality, dronified law enforcement—but not fully predict. As President

Obama said, without irony, "From our use of drones to the detention of terrorist suspects, the decisions that we are making now will define the type of nation—and world—that we leave to our children."[14] This existential turbulence batters the Predator Empire, engulfing it in vortexes that hurl it toward unexpected horizons of human coexistence.

The applications of aerial surveillance are limitless. As discussed in chapter 5, a dronified police force could soon saturate the city with swarms of autonomous microdrones invading the privacy of its newly exposed residents. Indeed, the future of U.S. policing is of particular concern, since the war machine and its corporate lobbyists are pushing for dronified law enforcement (the U.S.–Mexico borderlands are already a case in point). Imagine, then, a panopticon, or a dronopticon, that resembles not a prison but a flock of birds: a swarm that effectively encloses hundreds, thousands, perhaps even one day millions of people inside dronespheres. The psychological costs we will pay—paranoia, fear, anxiety—for living inside these atmospheric prisons will be excruciating, as many of the residents of Pakistan's tribal areas have experienced for over a decade. An imminent existential danger posed by the Predator Empire is this atmospheric cage, one in which thoughts, cognitions, and affects are reengineered by the material infrastructures of a dronified surveillance state—the new normal. Or put another way, the Predator Empire is not only producing new geopolitical spaces across the planet but also creating new cognitive spaces: existential shells for humanity to live and die inside. These mobile bubbles of state surveillance are rapidly foaming in the domestic landscapes of technological civilization.

The hazard posed by dronified policing and dronified killing is well documented. "The danger here is relinquishing the state's sovereign power to kill," argues Stephen Graham, "and delegating it to assemblages of silicon, titanium and software code—to perform acts of killing which not only are unbound from the defined times and spaces of traditional wars but also fall conveniently distant from the capricious gaze of mainstream media."[15] But this existential emergency, as important as it is, is just the tip of a much more systematic danger: the rule by Nobody.

Arendt, to recall, feared bureaucratic forms of control were creating an oppressive and unaccountable form of state power. As she describes it, "Bureaucracy is the form of government in which everybody is deprived of political freedom, of the power to act; for the rule by Nobody is not no-rule, and where all are equally powerless we have a tyranny without a tyrant."[16] If power is upheld by legitimate political institutions—the living power of people interacting with each other—then the rule by Nobody is conversely a nonliving power, a petrified state apparatus of force. Speaking of the Leviathan, Latour and Callon write, "There is no overall architect to guide it, and no design, however unreflected."[17] The rule by Nobody is the Leviathan shorn of living flesh, a hideous skeleton in which computers think, machines act, and drones kill.

Indeed, with the ascendency of drone warfare, we can see the most sophisticated (and robotic) rule by Nobody so far. Arendt, in a quote worth repeating from chapter 4, writes, "Only the development of robot soldiers, which, as previously mentioned, would eliminate the human factor completely and, conceivably, permit one man with a push button to destroy whomever he pleased, could change this fundamental ascendancy of power over violence."[18] With this sentence penned half a century ago, did Arendt not predict the birth of the Predator Empire? The rise of violence over power? The rise of machines over marines? In this sense a dronified state apparatus not only mediates the practice of state power but also reconfigures the very interface between the state and the population. Consequently, the war of all against all runs as the background condition of state power, since machines endlessly attack and reconfigure human existence. As Latour concludes, "*We have never left the state of war*, the *state of nature* that Hobbes thought the Leviathan had gotten us out of,"[19] because a war of technological mediation endlessly translates the worlds in which humans live and die.

Fearing the rise of nuclear weapons during the Cold War, Arendt warns, "The seemingly irresistible proliferation of techniques and machines, far from only threatening certain classes with unemployment, menaces the existence of whole nations and conceivably of all mankind."[20] The kind of eradication augured by the Predator Empire is not, however, the same as nuclear annihilation, but it is still an existential loss—a total war of a different kind. Instead of machines killing the entire species, what if the human condition becomes indistinguishable from the very infrastructures that enclose it? This, to recall, would be "a process of biological mutation in which human bodies gradually begin to be covered by shells of steel. For the watcher from the universe, this mutation would be no more or less mysterious than the mutation which now goes on before our eyes in those small living organisms which we fought with antibiotics and which mysteriously have developed new strains to resist us."[21] This kind of alienation, in which the spaces of human dwelling and thinking become little more than artificial dronespheres, feeds directly into the conduct of state violence.

Crucial to the Predator Empire is this type of alienated violence. But it is not simply alienated in the way drone warfare is typically depicted (as a visual disconnection between pilot and battlefield). Instead, ever since President Nixon ended the draft in 1973 in the wake of the Vietnam War, state violence has become increasingly remote from an active public who are meant to legitimate the entire process. The electronic battlefield, tried and tested throughout Southeast Asia in the 1960s, slowly replaced the need for U.S. bodies in the kill chain: human senses, human limbs, and human decisions were outsourced to a growing prosthetic armada. As Gabriella Blum writes, "The end of conscription after Vietnam contributed to the notion of war as a government operation, much in the same way that operating prisons or maintaining seaports is a government function."[22] A growing fleet of drones means fewer American families see the effects of war.

In the space vacated by the engaged, active citizen stands the drone. "In this sense," argues Engelhardt, "think of us as moving from the citizen's army to a roboticized, and finally robot, military—to a military that is a foreign legion in the most basic sense. In other words, we are moving toward an ever greater outsourcing of war to things that cannot protest, cannot vote with their feet (or wings), and for whom there is no 'home front' or even a home at all. In a sense, we are, as we have been since 1973, heading for a form of war without anyone, citizen or otherwise, in the picture."[23] A rule by Nobody.

For domestic audiences, the war on terror is remote precisely in this sense: military violence is something that happens just out there—it's a government operation. Perhaps just as important, when human bodies are no longer pressed through the Predator Empire's mangle, fewer people care. Drone warfare is an illegitimate form of violence not principally because it breaks international law, as critics contend, or because it is morally abhorrent, but because it is severed from an active and engaged public. Dronified state violence has ascended over power. The mechanical monster does not need consent, a mass of people, or judicial forms of legitimation: it needs only instruments. In such times it becomes increasingly difficult—or simply unnecessary—to recruit soldiers to defend the state, thus legitimizing a growing mercenary and robot army. With fewer risks, dronified state violence becomes the increasingly default foreign policy option. Recall the 1971 testimony from Eric Herter, who spoke about the electronic battlefield in Vietnam. He warned of "an automated electronic and mechanical death machine whose killing will be one-sided, unseen and universal. . . . We have seen the mechanical monster, the mindless devastation, the agony of simple people caught in the fire storm of our technological rampage."[24]

For decades the U.S. homeland has been the victim of an intensifying and pathological form of immunization that understands security as the fossilization of (segregated) social relations and widespread obedience. This form of national security is enforced through the entrenchment of militarized policing and the rollout of a vast surveillance Panopticon. The health of communities is secondary to the feeding of the war machine. In other words, security is seen as a nonhuman rather than human condition. When the Leviathan no longer protects and cares for its people, a fundamental social contract is broken.

This broken contract in turn creates the conditions for the militarization of social life. Insecurity and alienation are highly productive—and profitable—forces for the mass enclosure of humanity. "The extreme form of power is All against One," argues Arendt, "the extreme form of violence is One against All. And this latter is never possible without instruments."[25] Power belongs to the people, the multitude, the All. Without this congregation to support it, the sovereign—the One—fearing its power slipping away, will impose its rule with force, with violence, against the All.

In other words, as long as people are a part of the Leviathan, a source of instability, resistance—even redirection—is possible. But the robotic Leviathan, like the robotizing technological civilization from which it emerges, continually expels living flesh, generating an existential contradiction at the heart of the international system. The impulse to replace power with violence has been a historically consistent trend and typically characterizes the death rattle of empires. When all bonds are broken, when society is fully enclosed, a form of terror is unleashed upon the population. At its highest stage, the Leviathan transforms into a war machine, and the national security state conquers the social security state. We die so that the Predator Empire may live.

Lone Wolves

A hallmark of modernity is located in the assumption war takes place between nation-states and their respective militaries. This no longer describes the geography of contemporary world violence, if it ever did. Indeed, the Hobbesian state of nature—of violence targeted against, through, and between individuals—has returned as a central organizing principle of the Predator Empire. Fear of the lone wolf has come to symbolize this transformation in the United States. As President Obama warned, "Deranged or alienated individuals—often U.S. citizens or legal residents—can do enormous damage, particularly when inspired by larger notions of violent jihad."[26] Former U.S. attorney general Eric Holder reiterated the same point, "The thing that I think keeps me up most at night [is] this concern about the lone wolf who goes undetected."[27] Hunting dangerous individuals, at home and abroad, is fast becoming the center of U.S. national security strategy. "The barbarity of war between states," argues Mary Kaldor, "may have become a thing of the past. In its place is a new type of organized violence that is more pervasive and long-lasting."[28] Targeted killing has come to crystallize what Blum calls the "individualization of warfare."

On the face of it, the individualization of warfare represents the lesser of two evils: a form of military violence that embodies an important transition away from the kind of indiscriminate mass killing that tore through the twentieth century. As Blum writes, "As the figure of the individual loomed larger, nationality and geography have become increasingly overshadowed by universal commitments to the equal worth and human dignity of each individual life."[29] An appeal is being made—at least on paper—to a more generic humanity, beyond the symbolic housing of any national community. However much the humanitarian impulse of targeted killing is celebrated, it is accompanied by its more troubling obverse: the right to intervene, the right to kill. By celebrating the cosmopolitan ideals of individualized killing, one is celebrating not pacifism but efficient killing. The Predator Empire dispatches Reapers and not Angels after all.

To recall Grégoire Chamayou's argument, "What is emerging is the idea of an invasive power based not so much on the rights of conquest as on the rights of pursuit: a right of universal intrusion."[30] In this way the Predator Empire simultaneously defends and attacks the international system. In order to preserve the world system of pacified, civilizatory enclosures, U.S. drones have frequently violated the very thing they seek to defend. As Stuart Elden argues, "Territorial integrity as preservation wins out over territorial integrity as sovereignty."[31] The war on terror as a war against the individual thus produces a geography of violence that spirals beyond national borders. By hunting individuals, the space of the target simultaneously expands and contracts, moving between a global Panopticon and the individual prey. Predator drones, perhaps more than any technology, have embodied, enabled, and executed this form of predation.

By hunting individuals, state violence comes to resemble a global policing operation. In this sense the Predator Empire does not end the Hobbesian war of all against all but institutionalizes an atomized system of worldly violence. As Michael Hardt and Antonio Negri make clear, "We have entered the era of minor and internal conflicts. Every imperial war is a civil war, a police action—from Los Angeles and Granada to Mogadishu and Sarajevo."[32] Replacing the idea of a war between nations is this idea of a universalized neo-Hobbesian state of nature in which the sovereign hunts down individuals. Enemies of the state are not based on a national status but on their perceived or actual threat. Death warrants, signed by Nobody, can be dispatched to anybody.

The war on terror certainly began with more defined territorial registers: the invasion and occupation of Afghanistan and Iraq. As U.S. national security strategy shifted away from this kind of large-scale counterinsurgency, however, it left exposed the naked ambition to kill enemies of the state wherever they appeared—a geopolitics based solely on the delivery of death. This type of conflict works beyond conventional spatial containers of violence, shifting from a discrete battlefield to an amorphous battlespace, or what Derek Gregory has called an "everywhere war."[33] The historical manifestation of total war—a mass of bodies subject to industrial slaughter—is morphing into another type of total war, based on a universalized policing of the social field.

By reengineering the war on terror as a global policing operation, U.S. citizens are increasingly placed under the same set of expectations, rights, and responsibilities as individuals outside the homeland. Across many parts of the United States, neighborhoods are regularly treated as battlefields, and criminals, as enemy combatants. The war on terror has consistently internalized the existence of the enemy within its borders. In the wake of the 2013 Boston Marathon bombings, for example, prominent Republican Lindsey Graham commented, “This is Exhibit A of why the homeland is the battlefield,” adding, “It sure would be nice to have a drone up there.”[34] Of course, many police forces do have drones in the skies patrolling the homeland, just as U.S. Customs and Border Protection fly Predators along the U.S.–Mexico border. Eric Holder, when pressed about the use of drones to strike U.S. citizens on American soil, admitted that in times of “emergency” the president could “authorize the military to use such force if necessary to protect the homeland.”[35]

The opposing but interlocking figures of the lone wolf and the Predator drone have become symbolic of the domestication of the war on terror. A planet constantly at war—a clash between wolves and Predators—infiltrates and degrades some of the most intimate spaces of everyday life. There's no denying that lone wolves and small terrorist cells pose a real threat, as the 2015 attacks in Paris demonstrated so viscerally. They're just not a particularly big threat—and perhaps nowhere near as threatening as the surveillance apparatuses the Predator Empire installs to track them down. As Michael Harwood observes, “With shepherds like these guarding the flock, wolves may be beside the point.”[36] Moreover, in some cases, these lone wolves are products of the system itself. As the national security state replaces the social security state, slowly eviscerating the livelihoods of millions, “it would,” argues Bernard Stiegler, “lead to a *state system of totalitarian terror, a politics of terror* to counter the terror of those in despair, with or without doctrines, isolated or in networks, operated by remote control or emerging spontaneously as a horrifying new generation . . . *and all suicidal in some individual or collective way.*”[37]

Under such an understanding, the war on terror confronts those outbursts of subjective nihilism with a systematic nihilism. Is it any wonder the Predator Empire faces down an individualized threat when more and more individuals are expelled from the very protections once celebrated, and guaranteed, by technological civilization? Increasingly, then, the modern subject finds itself trapped between two nihilisms, if we understand nihilism—through its Nietzschean inflection—as a process of exteriorization, of emptying and evacuating the world of its meaning, substance, and durability. A suicidal terrorism clashes with a self-destructive form of capitalist technics, both locked in deadly embrace. Speaking to the latter, Alain Badiou argues, “Capitalist nihilism has arrived at a stage of the non-existence of any world. Yes, today there is no world, there is nothing but a group of singular disconnected situations. . . . Today, outside of the grand and petty bourgeoisie of the imperial cities, who proclaim themselves to be ‘civilization,’ you have nothing apart from the anonymous and the excluded.”[38] The task of philosophy, of an emancipatory thinking, must be to escape this synthesis of two nihilisms. So many centuries have elapsed since the enclosure of the commons, and yet we still remain strangers in a strange land.

Lonely Wolves

There is a profound relationship, synergy even, between the atomization of technological civilization and the individualization of the Predator Empire. The disenchantment of the world and the rise of the individual as the locus of state violence feed into each other. The passage from the social security state to a national security state is felt deep in people’s bodies and minds. Legions of psychologically disaffected and physically impoverished people fuel the Predator Empire. Globalization, in particular, has universalized the atomization of the species, working to connect individuals to each other, but as hyperindividualized units rather than as unified conglomerates. What is paradoxical about contemporary globalization, then, is it unites people, but in a very disjunctive arrangement: a synthesis, yes, but a disjunctive synthesis—a connective disconnection.

Humanity did not find itself in the world interior of technological civilization overnight, of course. The old spaces of human coexistence and political subjectivity were enclosed over centuries. A key date is 1492, after the so-called New World was “discovered” by Christopher Columbus. This began the dawn of European colonization and the rise of transatlantic powers, including the Portuguese, Spanish, and British empires. In the wake of this seafaring age, oceans and mountains, which had previously divided the world’s occupants, struggled to incubate a patchwork of unique semiospheres in the face of a universalizing imperial space.

Humanity found itself cast adrift, torn from the intimate constellations of soil, sea, and sky. Local immune structures, the cultural and religious baldachins of dwelling, were disenchanted, sometimes violently, by the homogenizing effects of geometry and capital. Globalization, argues Peter Sloterdijk, "bursts open the dream shells of grounded, housed, internally oriented and autonomously salvific collective life."[39] Imperial space, the space of empire, leveled everything to a common denominator of indifferent economic calculation. The planet was deterritorialized—its rich collection of culturally unique dream shells were torn open by empire. A globalized geoeconomic colonization connected both sides of the Atlantic, and capital reformatted disparate lifeworlds into a mesh of unrelated "locations."

From this newly imagined tabula rasa, a new congregation of immunitary structures was subsequently installed. The nation-state, following the symbolic 1648 Treaties of Westphalia, came to organize and house vast blocs of humanity inside its territorial enclosures. The planet was patterned with these real and imaginary Leviathans, each producing insides and outsides, friends and enemies.

But this topography could not possibly withstand the universalizing imperatives of globalization. "What the sixteenth century set in motion was perfect by the twentieth: no point on the earth's surface, once money had stopped off there, could escape the fate of becoming a location."[40] Nations interpenetrated each other with capital and commodities, data and disease. The species became entangled as never before, even if it was a connection born of a universal disconnection. The telecommunications revolution in turn shattered the ancient regime of face-to-face human contact. Electronic globalization announced itself with radio waves bouncing through the atmosphere, satellites orbiting in space, and fiber-optic cables snaking beneath the seas. Technological civilization came to unite us inside a growing electronic atmosphere, and these existential infrastructures power the Predator Empire.

As masses of shocked, displaced, and alienated people were brought into the interior of the capitalist system, beginning with English enclosure, an increasingly individualized society was manufactured. From this pool of insecurity, a proliferation of security and surveillance apparatuses would spawn, setting in motion the great age of confinement. As Friedrich Engels argues, "Present-day society, which breeds hostility between the individual man and everyone else, thus produces a social war of all against all."[41]

In short, as technological civilization unified the world's population in densely arranged enclosures, it came to realize more and more effectively what Engels calls a society "composed wholly of atoms." The Hobbesian war of all against all rather than being dispelled by a new economic nirvana was being actively engineered by modern capitalist forms of sovereignty that pitted individuals against each other. As Sloterdijk explains, "If classic modernism banked on immunization through the collectivity . . . the construction of postmodern immune systems puts more of an accent on individuals."[42]

The vast metropolises of technological civilization began to glue humanity in acrimonious and estranged densities. This disjunctive synthesis in turn "creates a situation of permanent social danger and requires the powerful apparatuses of the society of control to ensure separation and guarantee the new management of social space."[43] We may very well be connected in a globalized world, but our fusion, our coming together, is of an extremely individualized form. Here, Sloterdijk uses the metaphor of "foam" to describe the proliferation of such fragmented spaces: "What is currently being confusedly proclaimed in all the media as *the* globalization of the world is, in morphological terms, the universalized war of foams."[44] Compact cohabitation should not therefore be mistaken for companionship, since a systematic loneliness grips technological civilization. The social war of all against all is expressed through a capsular, fragmented, and disjunctive spatiality.

The commons, while certainly no utopia, sustained a vital link between earth and human being. The social war of all against all, mobilized during the enclosure of the commons, pierced the protective existential shells that once housed villages, towns, and communities in autonomous lifeworlds. Enclosure left in its wake a deworlded and alienated humanity. "The loss of the commons entailed, for the poor, a radical sense of displacement."[45] As E. P. Thompson continues, "The social violence of enclosure consisted precisely in the drastic, total imposition upon the village of capitalist property-definitions."[46] Communal spheres of coexistence were remade into neurotic foams of dissociated individuals. The atomization of humanity produces endless contradiction throughout technological civilization. As Hardt and Negri write, "The problem of imperial administration is thus to manage this process of integration and therefore to pacify, mobilize, and control the separated and segmented social forces."[47]

For a long time religion was used to pacify these segmented social forces. The transcendental order for domesticating humanity—the theological matrix, the divine order—began to flicker and, eventually, to fade during the modern age's disenchantment. "It is this tear," explains Roberto Esposito, "that suddenly opens in the middle of the last millennium in the earlier immunitarian wrapping that determines the need for a different apparatus of the artificial sort that can protect a world that is constitutively exposed to risk."[48]

If the modern age killed its gods, then technological civilization had to replace them with a form of supreme sovereignty no less divine, and no less powerful. "In the course of history," argues Jacques Ellul, "there have always been different principles of civilization according to regions, nations, and continents. But today everything tends to align itself on technical principles."[49] Machines rather than deities now domesticate masses of individuals, herding them inside the glowing interior of technological civilization. The Predator Empire in turn polices these civilizatory enclosures, keeping us on the great inside. Of course, the drone may bring people on the great inside, but that is not to suggest comfort or protection, only surveillance, or what Ariel Handel calls "exclusionary surveillance."[50] Most people on the inside still exist as outsiders: one can be surveilled and abandoned at the same time.

In the so-called badlands of the planet, U.S. drones colonize technological civilization's rebellious frontiers with ever more advanced forms of necropower. As Achille Mbembe explains, these "deathworlds" are "new and unique forms of social existence in which vast populations are subject to conditions of life conferring upon them the status of *living dead*."[51] The Predator Empire does not occupy these spaces directly but enforces its domination from above. In this sense the dronification of state violence embodies a form of nonterritorial occupation. Do we not here witness a nightmare planet torn apart by overlapping spaces of hyperenclosure and exception, of imprisoned living and extrajudicial death? In other words, enclosure is not simply a power to preserve or immunize life but an exceptional power to take life. This topology functions everywhere, even across U.S. streets, which have come to mirror the tortured geographies of occupied cities abroad.

The Drone

The desire to enclose the world in a single immunitary configuration may be as old as empire, but the drone is a technology that can begin to realize this ambition. In the Predator drone, the U.S. military discovered an access weapon that enabled it to police and surveil the planet, exposing and pacifying restive populations wherever they surface. "War," writes Engelhardt, "has always been a quintessentially human and yet inhuman activity. Now, it seems, its inhuman aspect is on the rise. With the US military working to roboticize the future battlefield, the American way of war is destined to be imbued with *Terminator*-style terror."[52] Predator drones have transformed the U.S. immunity system by bringing distant spaces into the gaze of pilots, targeters, and algorithms thousands of miles away. As Eyal Weizman argues, the Orient no longer occupies a space in the East, across oceans and mountains, "but vertically under the tyranny of a western airborne civilization manning, or remotely manning, armed platforms above it."[53] Securing the skies with blimps, planes, and drones has continually reengineered the relationships between space, power, and sovereignty. "This technological development," writes Paul Virilio, "has carried us into a realm of factitious topology in which all the surfaces of the globe are directly present to one another."[54]

The Predator Empire marks the evolution of U.S. military power away from a boots-on-the-ground model of counterinsurgency enabled by a topography of supersized bases to a dronified model of counterterrorism enabled by the topological exchange of computer data. As Sloterdijk argues, "Virtual shells have replaced the imagined ethereal sky; thanks to radio-electronic systems, the meaning of distances has effectively been negated in the centers of power and consumption. The global players live in a world without gaps."[55] The Predator Empire is productive of a globalized ecumene in which space is erased by technics, topography is erased by topology, the ground is erased by the sky, and the present is erased by the future. It is the electronic battlespace of the twenty-first century. While by no means denying the vast material infrastructure of Droneworld that coordinates targeted killings across the globe, the extensive digitizing, coding, and remote elimination of life in real time are what mark the Predator Empire as biopolitically and geopolitically distinctive.

The Predator drone extends its policing power to how war looks and feels. As Virilio famously argues, "There is no war, then, without representation."[56] However, this formulation can be understood in reverse: there is no representation without war, which is to say, there is no image without struggle. The drone needs to be considered in this phenomenological light, stamping out bits and pieces of reality like a jigsaw puzzle. There is a technical conflict over who is imaged, what is imaged, and how they are imaged. The landscapes and spaces of human coexistence processed by the drone become contaminated with a technological logic that seeks to sanitize the practice of remote killing. Drones are not simply instruments of a superior human intentionality, then, but also responsible for policing the aesthetic conditions of state violence. For Virilio the stakes are clear: "The intensity of automatic weaponry and the new capacities of photographic equipment combine to project *a final image of the world,* a world in the throes of dematerialization and eventual total disintegration."[57]

In Guy Debord's *The Society of the Spectacle,* the philosopher argues we live in a society of hypermediated visual reality. What was once directly lived has receded into representation: "Spectators are linked solely by their one-way relationship to the very center that keeps them isolated from each other. The spectacle thus reunites the separated, but it reunites them only *in their separateness.*"[58] The spectacle, in other words, concretizes a set of social relations based on a disjunctive synthesis. The drone is perhaps the most refined weapon of the society of the spectacle, since it redistributes the spectacularity of the world, policing the abstract system of what is seen, heard, and felt. The Predator Empire thus tears open the planet only to enclose it within a vast archipelago of electromagnetic holograms, producing cinematic warspheres that saturate and invert reality. We need to imagine this aesthetic policing in its totality—beyond the visual—to encompass the feelings, emotions, and thoughts of those surveilled populations. The drone, the latest in a long line of civilizatory apparatuses, acts like a virus, boring into the civilizatory shells of human beings and reprogramming the climate of their interiors.

An apparatus, according to Giorgio Agamben, is "literally anything that has in some way the capacity to capture, orient, determine, intercept, model, control, or secure the gestures, behaviors, opinions, or discourses of living beings."[59] Much like Gilbert Simondon and Bernard Stiegler, Agamben argues the modern individual emerges at the interstice of life and technology: "Apparatus, then, is first of all a machine that produces subjectifications, and only as such is it also a machine of governance."[60] Humanity has for millennia sought to domesticate its most dangerous instincts with apparatuses, bringing its destructive powers inside the great civilizatory spheres. But the protection offered by these apparatuses can be corrosive. The problem here is that the subject not only is the subject of an apparatus (i.e., its master) but is simultaneously subjected to the apparatus. The modern individual has become imprisoned by a technological civilization that seeks to secure and police everything.

Ultimately, then, drones not only patrol the skies of a world but simultaneously police its existential atmospheres. Drones are world-making and world-destroying weapons. The Predator Empire is not simply a geopolitical power but also an existential and spectacular power, one that probes deep into the minds and bodies of those caught inside its security spheres. This existential feeling, more than ever, is defined by a pervasive anxiety. In Pakistan, a repeated site of targeted killings, a widespread psychological damage has already been internalized by many communities. To recall one journalist's testimony, "If I am walking in the market, I have this fear that maybe the person walking next to me is going to be a target of the drone. If I'm shopping, I'm really careful and scared. If I'm standing on the road and there is a car parked next to me, I never know if that is going to be the target."[61] Aerial assassination changes people's sleeping patterns, daily behavior, and even friendship circles. The biopolitical logic of drone strikes is not just death, then, but the ordering and policing of the lifeworld. The Predator Empire, in short, fights to control the very contours of existence. Across the planetary spectrum, the spaces of life and liberty are being strangled by a unifying civilizatory battlespace. As Chamayou writes, "Drones are indeed petrifying. They inflict mass terror upon entire populations. It is this—over and above the deaths, the injuries, the destruction, the anger, and the grieving—that is the effect of permanent lethal surveillance: it amounts to a psychic imprisonment within a perimeter no longer defined by bars, barriers, and walls, but by the endless circling of flying watchtowers above."[62]

If we come to understand our lives as inseparable from the spheres in which we dwell, perhaps we'd take more seriously the toxicity of this imprisonment. The atmosphere is not a neutral background to human life but the space of subjectivity. If the Predator Empire successfully secures the interiors of human existence, modulating its psychosocial atmospheres, then it polices the very grammar of our souls. There is no eternal, obdurate human essence that can easily resist the individuation of imperial technics. World War III—the omnicrisis of today, the globalized social war—is fundamentally an existential war. The Predator Empire targets those ecosystems in which individuals live, become, and die. For this reason the existential crisis we confront with the Predator Empire is fundamentally spatial and requires a more-than-human geopolitics to fully tease out the consequences and scale of an expanding Droneworld. Moreover, this takes us away from the prevalent Cartesian reification that invests too much authority in state leaders, mystifying the nonhuman power of the Leviathan.

The End

The dream of technological civilization, of a single artificial paradise that cocoons humanity from the frost outside, is illusory. Step beyond the spectacular glow of its exterior, and you'll discover the glittering façade was hiding an ugly apartheid. In technological civilization we are brought together only to be cast apart in splintering, suffocating spheres. This is how globalization must be understood—as a disjunctive synthesis, a *universalizing apartheid.* For centuries this social war of all against all has enclosed the commons and placed humanity under surveillance. This "eclipse of a common public world," warns Arendt, is "so dangerous in the formation of the worldless mentality of modern ideological mass movements."[63] In Arendt's analysis totalitarian systems parasite on atomized individuals. At its most depraved, this form of total domination "strives to organize the infinite plurality and differentiation of human beings as if all humanity were just one individual."[64]

The atomization of the species has only accelerated since Arendt warned of its perils in the wake of the Nazi regime, which means the totalitarian impulse remains. The existential conditions are certainly present: a planet inhabited by billions of people who are subjected to a civilizatory system that, increasingly, no longer protects them but only polices them: targeted by a Leviathan reborn as a war machine. In such cases the war on terror is transforming into a forever war against civilizatory outsiders and the surplus populations that can be found everywhere. In many ways this brave new world is already here. But its spatial form of control has changed from the Gulags and concentration camps of the twentieth century. Totalitarianism now resembles an open prison—or a dronopticon—where the spaces of everyday life are captured in an electromagnetic surveillance regime.

The totalitarianism we face today is the Predator Empire, a form of sovereignty that is technologically rather than territorially based.

For decades, millennia even, sovereignty has required that the will of the multitude be transferred to a monarch, emperor, or some form of elected parliament. This movement nonetheless represents a type of worldly alienation, since that which is preserved—human life and freedom—is transferred to a higher authority. In the act of self-preservation, then, life is pulled beyond its own fleshy boundaries—is exteriorized—in the search for a transcendental refuge in which it may be secured. Hobbes argues this refuge is, of course, the Leviathan. The problem is this Leviathan is not—and never was—simply a human body. It has always been more-than-human, since the Leviathan is constructed with a range of objects, materials, tools, weapons, infrastructures, and apparatuses. Biological insecurity is continually traded for artificial security. As Stiegler explains, "In other words, the *question of war* is inevitably contained within the *question of technics*: the technical tool is above all an organ of predation and defence. Technicity, as a system, contains the artificial and social system of predation and defence from the beginning of humanity."[65]

This means the raw materials of sovereignty (and violence) are in technical and historic flux. In the age of technological civilization, the Leviathans spawned from the bowels of a hypermediated reality have come to resemble mechanical monsters, increasingly shorn of their human qualities, aspirations, and emotions. When the sovereign that rules over a commonwealth is Nobody, a systematic crisis threatens to tip the balance of violence over power. "No government exclusively based on the means of violence has ever existed," argues Arendt.[66] Yet with the enlargement of the national security state at the expense of the social security state, that's exactly the danger we face. This rule by Nobody would be the installation of a robotic and unaccountable government utterly indifferent to the health of the commonwealth—the ultimate expression of a Hobbesian artificial person.

The war of all against all is not displaced by sovereignty, then, but represents a more pervasive process of mediation in which the human condition is assailed by artificial methods of control. In other words, individuation, the physical and psychological production of individuals, "is a state of permanent war."[67] Every day, knowingly or not, we wake up in the middle of this struggle. Since we live in a state of continual exteriorization, our very humanity is remade by the objects we dwell alongside. As Arendt insists, "If the human condition consists in man's being a conditioned being for whom everything, given or man-made, immediately becomes a condition of his further existence, then man 'adjusted' himself to an environment of machines the moment he designed them."[68] Our individuation passes through the synthetic organs of technological civilization, which means we are always a kind of transindividual formed across technical, ecological, and psychological forces. We never fail to pass through a planet constituted by shifting technogeographies, and in each one of these augmented, spheric encasings, we are conditioned by its nonhuman occupants and elements. The more one searches for an invariant, eternal human form, the more one discovers a succession of cyborgs, a parade of tool-beings who are beholden to the technical artifacts they created. We are wordsmiths, no doubt, but domination belongs to the blacksmiths of the world: those women, men, and machines that engineer our sociotechnological systems. As Hobbes writes, "And covenants without the sword are but words, and of no strength to secure a man at all."[69]

If we exist in augmented shells, then consider the Predator Empire a universalizing shell shock: a rapidly expanding form of trauma tearing through the control societies of technological civilization. We exist, according to Michel Foucault, "in the panoptic machine, invested by its effects of power, which we bring to ourselves since we are part of its mechanism."[70] This kind of society realizes a militaristic dream to surveil and control everything, to produce human beings who are "meticulously subordinated cogs of a machine."[71] Recall the NSA engineer who wrote the 1971 paper "Crime Deterrent Transponder System," which contains the designs for converting the city into a gigantic sensor prison. This artificial dome engulfs "the criminal with a kind of externalized conscience—an electronic substitute for the social conditioning, group pressures, and inner motivation which most of the society lives with."[72] The problematic posed by the Predator Empire thus far exceeds the dronification of state violence. The greater danger lies in the dronification of the human condition: the mass production of anxious, paranoid, highly atomized individuals secured in their comfort capsules, soothed—and distressed—by the buzz of police drones stalking the skies.

The question is not simply whether we are the masters or slaves of the coming drone army but whether drones create a better world for us to inhabit. Already, we are housed beneath the electronic skin of technological civilization. Humanity, writes Sloterdijk, creates its own worlds "on artificial continents under artificial skies and domes."[73] This book opens with the curious case of the Winooski dome: a blueprint for a town that never was. "I'm not a sociologist," to recall the mayor in 1981, "but the idea of people living together in a controlled environment is a much more complex question than any of the technical concerns."[74] Although the idea of wrapping a town in a plastic bubble seems far fetched, the logic of enclosure—of protecting life from the outside—is not reducible to physical shapes. Enclosure no longer needs concrete barriers to create pacified atmospheres. Instead, aerial technologies can pervade the lifeworld with an equally totalizing effect, creating securitized spheres of human activity.

If domes represent forms of atmospheric control—in which human lives are contained inside suffocating, sometimes intangible carceral husks—could the entire planet one day be contained by a single synthetic orb? Such a vast enclosure, in which our lives are constantly policed, expresses a militaristic dream of full spectrum dominance: a terrordome. The Reagan-era Star Wars space shield was an early attempt to immunize the entire planet, to place Earth inside a bubble of U.S. military power. But the Predator Empire of today goes further by going smaller. Drones are the new saviors of our neurotic anthropology. If the godly heavens above once protected us from the demons below, then consider the Predator Empire as our new civilizatory ceiling, enclosing humanity in the great robotic inside—the best of us, the worst of us, and the last of us.

Notes

Introduction

1. David Appell, "Doomed Dome: The Future That Never Was," *Humanity+,* September 30, 2009, www.hplusmagazine.com.

2. Matthew Vita, "Winooski, Vt., Lobbies for a Dome over the Town," *Free Lance-Star,* March 27, 1980.

3. Editorial, "The Winooski Dome," *Kentucky New Era,* November 24, 1979.

4. Vita, "Winooski, Vt., Lobbies for a Dome over the Town."

5. Hannah Arendt, *The Human Condition,* 2nd ed. (Chicago: University of Chicago Press, 2013), 9.

6. Peter Sloterdijk, *Spheres,* vol. 1, *Bubbles,* trans. Wieland Hoban (Los Angeles: Semiotext[e], 2011), 28.

7. Judith Butler, *Frames of War: When Is Life Grievable?* (London: Verso, 2009), 23.

8. Associated Press, "Vermont Town's Dome Dream Dims Like Words to Anthem," *Toledo Blade,* October 12, 1981.

9. Arendt, *The Human Condition,* 147.

10. A slight disclaimer is needed at this stage. With a topic like drone warfare, indeed any kind of conflict, many of the empirical details, facts, and other pieces of information presented in this book may have changed considerably since I finished writing the bulk of the manuscript at the start of 2015.

11. Tom Engelhardt, "The Fourth Branch: The Rise to Power of the National Security State," *TomDispatch,* August 3, 2014, www.tomdispatch.com.

12. Sloterdijk, *Spheres,* 65.

13. Peter Linebaugh, *Stop, Thief! The Commons, Enclosures, and Resistance* (Oakland, Calif.: PM Press, 2014), 38.

14. "Risk of Robot Uprising Wiping Out Human Race to Be Studied," BBC, November 26, 2012, www.bbc.co.uk/news.

15. Human Rights Watch and International Human Rights Clinic, *Losing Humanity: The Case against Killer Robots* (November 2012).

16. Rory Cellan-Jones, "Stephen Hawking Warns Artificial Intelligence Could End Mankind," BBC News, December 2, 2014, www.bbc.co.uk/news.

17. Ian G. R. Shaw and Majed Akhter, "The Dronification of State Violence," *Critical Asian Studies* 46, no. 2 (2013): 211–34.

18. Jeremiah Gertler, *U. S. Unmanned Aerial Systems* (Washington, D.C.: Congressional Research Service, January 2012).

19. Micah Zenko, "Obama's Embrace of Drone Strikes Will Be a Lasting Legacy," *New York Times,* January 12, 2016, www.nytimes.com.

20. "U.S. to Cut Almost 100,000 Troops," BBC News, January 26, 2012, www.bbc.co.uk/news.

21. David E. Sanger, *Confront and Conceal: Obama's Secret Wars and Surprising Use of American Power* (New York: Crown Publishers, 2012), 270.

22. Alfred W. McCoy, "Imperial Illusions: Information Infrastructure and the Future of U.S. Global Power," in *Endless Empire: Spain's Retreat, Europe's Eclipse, America's Decline,* ed. Alfred W. McCoy, Josep M. Fradera, and Stephen Jacobson (Madison: University of Wisconsin Press, 2012), 384–85.

23. Nick Turse, *The Changing Face of Empire: Special Ops, Drones, Spies, Proxy Fighters, Secret Bases, and Cyberwarfare* (Chicago: Haymarket Books, 2012), 30.

24. Graham Harman, *Tool-Being: Heidegger and the Metaphysics of Objects* (Chicago: Open Court Publishing, 2002), 212.

25. Jacques Ellul, *The Technological Society,* trans. John Wilkinson, (New York: Vintage Books, 1964), 276.

26. Peter W. Singer, *Wired for War: The Robotics Revolution and Conflict in the 21st Century* (New York: Penguin, 2009), 318.

27. Graham Harman, *Bruno Latour: Reassembling the Political* (London: Pluto Press, 2014), 18.

28. Gabrielle Hecht, introduction to *Entangled Geographies: Empire and Technopolitics in the Global Cold War,* ed. Gabrielle Hecht (Cambridge, Mass.: MIT Press, 2011), 3.

29. Alfred McCoy, "Fatal Florescence: Europe's Decolonization and America's Decline," in *Endless Empire,* ed. McCoy, Fradera, and Jacobson, 6.

30. Chalmers Johnson, *Blowback: The Costs and Consequences of American Empire,* 2nd ed. (London: Time Warner Paperbacks, 2002), 18.

31. Thomas Hobbes, *Leviathan: With Selected Variants from the Latin Edition of 1668,* ed. Edwin Curley (Indianapolis: Hackett Publishing Company, 1994).

32. Ibid., 76.

33. Peer Schouten, "The Materiality of State Failure: Social Contract Theory, Infrastructure, and Governmental Power in Congo," *Millennium: Journal of International Studies* 41, no. 3 (2013): 553–74.

34. Bruno Latour, *From Realpolitik to Dingpolitik or How to Make Things Public* (2005), 7, http://www.bruno-latour.fr/sites/default/files/96-DINGPOLITIK-GB.pdf.

35. Ibid, 5.

36. Nick Srnicek, "Representing Complexity: The Material Construction of World Politics" (PhD diss., London School of Economics, 2013), 42.

37. Ibid., 50.

38. Sallie A. Marston, John Paul Jones III, and Keith Woodward, "Human Geography without Scale," *Transactions of the Institute of British Geographers* 30, no. 4 (2005): 416–32.

39. Bruno Latour and Michael Callon, "Unscrewing the Big Leviathan: How Actors Macro-structure Reality and How Sociologists Help Them to Do So," in *Advances in Social Theory and Methodology: Toward an Integration of Micro and Macro-sociologies,* ed. Karin Knorr-Cetina and Aaron Victor Cicourel (Boston, Mass.: Routledge & Kegan Paul, 1981), 286.

40. Ibid., 284.

41. Ibid., 284.

42. Srnicek, *Representing Complexity,* 22.

43. Latour and Callon, *Unscrewing the Big Leviathan,* 294.

44. Schouten, *The Materiality of State Failure,* 560.

45. Bruno Latour, *Politics of Nature: How to Bring the Sciences into Democracy,* trans. Catherine Porter (Cambridge, Mass.: Harvard University Press, 2004), 218.

46. Andrew Barry, "The Translation Zone: Between Actor-Network Theory and International Relations," *Millennium: Journal of International Relations* 41, no. 3 (2013): 424.

47. Derek Gregory, "The Everywhere War," *Geographical Journal* 177, no. 3 (2011): 238–50.

48. Michael Hardt and Antonio Negri, *Empire* (Cambridge, Mass.: Harvard University Press, 2000), 189.

49. Mary Kaldor, *New and Old Wars: Organized Violence in a Global Era,* 3rd ed. (Cambridge: Polity Press, 2012).

50. Gregory, "The Everywhere War," 239.

51. Kaldor, *New and Old Wars,* 77.

52. Daniel Klaidman, *Kill or Capture: The War on Terror and the Soul of the Obama Presidency* (New York: Houghton Mifflin Harcourt, 2012), 201.

53. George W. Bush, "U.S. National Security Strategy: Strengthen Alliances to Defeat Global Terrorism and Work to Prevent Attacks against Us and Our Friends," U.S. Department of State Archive, September 14, 2001, http://2001–2009.state.gov/r/pa/ei/wh/15423.htm.

54. Sanger, *Confront and Conceal,* 252.

55. Ibid.

56. James W. Gibson, *The Perfect War: Technowar in Vietnam,* 2nd ed. (New York: Atlantic Monthly Press, 2000).

57. Paul Dickson, *The Electronic Battlefield* (Takoma Park, Md.: FoxAcre Press, 2012), 72.

58. Douglas Valentine, *The Phoenix Program* (Lincoln, Neb.: iUniverse, 2000), 420.

59. Alfred W. McCoy, "Confronting the CIA's Mind Maze," *TomDispatch,* June 7, 2009, www.tomdispatch.com.

60. Chalmers Johnson, "America's Empire of Bases," *TomDispatch,* January 15, 2004, www.tomdispatch.com.

61. Michel Foucault, *"Society Must Be Defended": Lectures at the Collège de France, 1975–1976,* trans. David Macey (London: Penguin, 2003), 240.

62. Roberto Esposito, *Bíos: Biopolitics and Philosophy,* trans. Timothy Campbell (Minneapolis: University of Minnesota Press, 2008).

63. Hannah Arendt, *On Violence* (New York: Houghton Mifflin Harcourt, 1969), 81.

64. Gilles Deleuze, "Postscript on the Societies of Control," *October* 59 (1992): 7.

65. Alfred W. McCoy, *Policing America's Empire: The United States, the Philippines, and the Rise of the Surveillance State* (Madison: University of Wisconsin Press, 2009), 19.

66. E. P. Thompson, *The Making of the English Working Class* (Harmondsworth, U.K.: Penguin Books, 1963), 238.

67. Foucault, *"Society Must Be Defended,"* 109.

68. Mike Davis, "The Urbanization of Empire: Megacities and the Laws of Chaos," *Social Text* 22, no. 4 (2004): 15.

69. Johnson, *Blowback,* 13.

1. The Long March to Human Enclosure

1. Peter Sloterdijk, *In the World Interior of Capital: Towards a Philosophical Theory of Globalization,* trans. Wieland Hoban (Cambridge, U.K.: Polity Press, 2013), 12.

2. Ibid., 139.

3. Peter Sloterdijk, *Spheres,* vol. 1, *Bubbles,* trans. Wieland Hoban (Los Angeles: Semiotext[e], 2011), 24.

4. Ibid., 25.

5. Ibid.

6. George W. Bush, "Transcript of President Bush's Address to a Joint Session of Congress on Thursday Night, September 20, 2001," CNN, September 21, 2001, http://edition.cnn.com.

7. Mark Neocleous, *War Power, Police Power* (Edinburgh: Edinburgh University Press, 2014), 137.

8. John Morrissey, "US Central Command and Liberal Imperial Reach: 'Shaping the Central Region for the 21st Century,'" *Geographical Journal* 182, no. 1 (2016): 15–26.

9. Michael Hardt and Antonio Negri, *Empire* (Cambridge, Mass.: Harvard University Press, 2000), 325.

10. Ibid., xi.

11. Peer Schouten, "The Materiality of State Failure: Social Contract

Theory, Infrastructure, and Governmental Power in Congo," *Millennium: Journal of International Studies* 41, no. 3 (2013): 563.

12. Graham Harman, *Bruno Latour: Reassembling the Political* (London: Pluto Press, 2014), 21–22.

13. Jacques Ellul, *The Technological Society,* trans. John Wilkinson (New York: Vintage Books, 1964), 128.

14. Ibid.

15. Agnes Heller, *A Theory of Modernity* (Malden, Mass.: Blackwell, 1999).

16. Ibid., 168.

17. Sigmund Freud, *Civilization and Its Discontents,* trans. Joan Riviere (Mansfield Centre, Conn.: Martino Publishing, 2010).

18. Ibid., 44.

19. Hardt and Negri, *Empire,* xii.

20. Simon Dalby, "Political Space: Autonomy, Liberalism, and Empire," *Alternatives: Global, Local, Political* 30, no. 4 (2005): 435.

21. John Agnew, "American Hegemony into American Empire? Lessons from the Invasion of Iraq," *Antipode* 35 (2003): 871–85.

22. Edward W. Said, "Representing the Colonized: Anthropology's Interlocutors," *Critical Inquiry* 15, no. 2 (1989): 218.

23. Chalmers Johnson, *The Sorrows of Empire: Militarism, Secrecy, and the End of the Republic* (London: Verso, 2004), 192.

24. Agnew, "American Hegemony into American Empire?," 883.

25. Andrew J. Bacevich, *Washington Rules: America's Path to Permanent War* (New York: Metropolitan Books, 2010), 20.

26. Chalmers Johnson, *Nemesis: The Last Days of the American Republic* (New York: Henry Holt, 2007), 19.

27. Mona Domosh, "Selling Civilization: Toward a Cultural Analysis of America's Economic Empire in the Late Nineteenth and Early Twentieth Centuries," *Transactions of the Institute of British Geographers* 29, no. 4 (2004): 453–67.

28. Hardt and Negri, *Empire,* 384.

29. Matthew Sparke, "American Empire and Globalisation: Postcolonial Speculations on Neocolonial Enframing," *Singapore Journal of Tropical Geography* 24, no. 3 (2003): 374.

30. Neil Smith, "The Endgame of Globalization" (presentation, Annual Elsevier Lecture in Political Geography, Association of American Geographers, Denver, Colo., April 5–9, 2005).

31. Ruth Oldenziel, "Islands: The United States as a Networked Empire," in *Entangled Geographies: Empire and Technopolitics in the Global Cold War,* ed. Gabrielle Hecht (Cambridge, Mass.: MIT Press, 2011), 16.

32. Gabrielle Hecht, introduction to *Entangled Geographies: Empire and Technopolitics in the Global Cold War,* ed. Gabrielle Hecht (Cambridge, Mass.: MIT Press, 2011), 2.

33. Oldenziel, "Islands," 32.

34. Chalmers Johnson, "America's Empire of Bases," *TomDispatch,* January 15, 2004, www.tomdispatch.com.

35. David Vine, "The Lily-Pad Strategy," *TomDispatch,* July 15, 2012, www.tomdispatch.com.

36. Sasha Davis, "The U.S. Military Base Network and Contemporary Colonialism: Power Projection, Resistance, and the Quest for Operational Unilateralism," *Political Geography* 30, no. 4 (2011): 220.

37. Tom Engelhardt, *The American Way of War: How Bush's Wars Became Obama's* (Chicago: Haymarket Books, 2010), 40.

38. Hannah Arendt, *The Human Condition,* 2nd ed. (Chicago: University of Chicago Press, 2013), 9.

39. Bruno Latour, *From Realpolitik to Dingpolitik or How to Make Things Public* (2005), 6, http://www.bruno-latour.fr/sites/default/files/96-DINGPOLITIK-GB.pdf.

40. Erika Cudworth and Stephen Hobden, "Of Parts and Wholes: International Relations beyond the Human," *Millennium: Journal of International Studies* 41, no. 3 (2013): 447.

41. Claudia Aradau, "Security That Matters: Critical Infrastructure and Objects of Protection," *Security Dialogue* 41, no. 5 (2010): 493.

42. Mike Bourne, "Guns Don't Kill People, Cyborgs Do: A Latourian Provocation for Transformatory Arms Control and Disarmament," *Global Change, Peace, and Security* 24, no. 1 (2012): 142.

43. Nick Srnicek, Maria Fotou, and Edmund Arghand, "Introduction: Materialism and World Politics," *Millennium: Journal of International Studies* 41, no. 3 (2013): 397.

44. Harman, *Bruno Latour: Reassembling the Political,* 35–36.

45. Nick Srnicek, "Representing Complexity: The Material Construction of World Politics" (PhD diss., London School of Economics, 2013), 59.

46. Cudworth and Hobden, "Of Parts and Wholes," 448.

47. Bourne, "Guns Don't Kill People," 157.

48. Karen Barad, "Posthumanist Performativity: Toward an Understanding of How Matter Comes to Matter," *Journal of Women in Culture and Society* 28, no. 3 (2003): 810.

49. Aradau, "Security That Matters," 494.

50. Barad, "Posthumanist Performativity," 821.

51. Charlie Savage and Mark Landler, "White House Defends Continuing U.S. Role in Libya Operation," *New York Times,* June 15, 2001, www.nytimes.com.

52. Peter W. Singer, "Do Drones Undermine Democracy?," *New York Times,* January 21, 2012, www.nytimes.com.

53. David Kennedy, *Of War and Law* (Princeton, N.J.: Princeton University Press, 2006), 96.

54. John O. Brennan, "The Ethics and Efficacy of the President's Counter-

terrorism Strategy" (remarks given at the Woodrow Wilson International Center for Scholars in Washington D.C., April 30, 2012).

55. Sarah Kreps and John Kaag, "The Use of Unmanned Aerial Vehicles in Contemporary Conflict: A Legal and Ethical Analysis," *Polity* 44: 260–85.

56. Grégoire Chamayou, *Drone Theory,* trans. Janet Loyd (New York: Penguin, 2015), 144.

57. Cudworth and Hobden, "Of Parts and Wholes," 449.

58. Caroline Holmqvist, "Undoing War: War Ontologies and the Materiality of Drone Warfare," *Millennium: Journal of International Studies* 41, no. 3 (2013): 535–52.

59. Ian G. R. Shaw and Majed Akhter, "The Unbearable Humanness of Drone Warfare in FATA, Pakistan," *Antipode* 44, no. 4 (2012): 1503.

60. Michel Foucault, *"Society Must Be Defended": Lectures at the Collège de France, 1975–1976,* trans. David Macey (London: Penguin, 2003), 244–45.

61. Sloterdijk, *In the World Interior of Capital,* 170.

62. Sloterdijk, *Spheres,* 84.

63. Ibid., 25.

64. Ibid., 24.

65. Arendt, *The Human Condition,* 52.

66. Sloterdijk, *In the World Interior of Capital,* 170.

67. Barad, "Posthumanist Performativity," 817

68. Giorgio Agamben, *What Is an Apparatus? And Other Essays,* trans. David Kishik and Stefan Pedatella (Stanford, Calif.: Stanford University Press, 2009), 15.

69. Peter Linebaugh, *Stop, Thief! The Commons, Enclosures, and Resistance* (Oakland, Calif.: PM Press, 2014), 142.

70. Stuart Hodkinson, "The New Urban Enclosures," *City: Analysis of Urban Trends, Culture, Theory, Policy, Action* 16, no. 5 (2012): 502.

71. Simon Fairlie, "A Short History of Enclosure in Britain," *The Land,* Summer 2009, http://www.thelandmagazine.org.uk/articles/short-history-enclosure-britain.

72. "Enclosing the Land," U.K. Parliament website, http://www.parliament.uk/about/living-heritage/transformingsociety/towncountry/landscape/overview/enclosingland/.

73. Karl Marx, *Capital: A Critique of Political Economy,* trans. Ben Fowkes (London: Penguin Books, 1990), 875.

74. Nicholas Blomley, "Making Private Property: Enclosure, Common Right, and the Work of Hedges," *Rural History* 18, no. 1 (2007): 9.

75. Ibid., 12.

76. Gerrard Winstanley, *The True Levelers Standard Advanced: Or, The State of Community Opened, and Presented to the Sons of Men,* (1649), https://www.marxists.org/reference/archive/winstanley/1649/levellers-standard.htm.

77. Hodkinson, "The New Urban Enclosures," 503.

78. David Harvey, *The New Imperialism* (Oxford: Oxford University Press, 2003).

79. Linebaugh, *Stop, Thief!,* 106.

80. "Introduction to the New Enclosures," Midnight Notes Collective website, 1990, http://www.midnightnotes.org/pdfnewenc1.pdf.

81. Alex Jeffrey, Colin McFarlane, and Alex Vasudevan, "Rethinking Enclosure: Space, Subjectivity, and the Commons," *Antipode* 44, no. 4 (2012): 1248.

82. E. P. Thompson, *Customs in Common* (London: Penguin Books, 1991), 165.

83. Marx, *Capital,* 915.

84. Achille Mbembe, "Necropolitics," trans. Libby Meintjes, *Public Culture* 15, no. 1 (2003): 26.

85. Peter Sloterdijk, "Foreword to the Theory of Spheres," in *Cosmograms,* ed. Melik Ohanian and Jean-Christophe Royoux (New York: Lukas and Stenberg, 2005), 227.

86. Thompson, *Customs in Common,* 170.

87. Linebaugh, *Stop, Thief!,* 2.

88. Mike Davis, *Planet of Slums* (London: Verso, 2006), 174.

89. Neocleous, *War Power, Police Power,* 74.

90. Hardt and Negri, *Empire,* 170.

91. Hodkinson, "The New Urban Enclosures," 509.

92. Linebaugh, *Stop, Thief!,* 157.

93. Félix Guattari, *The Three Ecologies,* trans. Ian Pindar and Paul Sutton (London: Athlone Press, 2000), 28.

94. Edward W. Soja, *Seeking Spatial Justice* (Minneapolis: University of Minnesota Press, 2010), 42.

95. Sloterdijk, *In The World Interior of Capital,* 193.

96. "Introduction to the New Enclosures," 5.

97. Hodkinson, "The New Urban Enclosures," 509.

98. Alain Badiou, "The Communist Hypothesis," *New Left Review* 49 (2008), www.newleftreview.org.

99. Slavoj Žižek, *Violence: Six Sideways Reflections* (London: Profile Books, 2008).

100. Jeffrey, McFarlane, and Vasudevan, "Rethinking Enclosure," 1259.

101. Blomley, "Making Private Property," 16.

102. Linebaugh, *Stop, Thief!,* 65.

103. Felix Driver, "Geography's Empire: Histories of Geographical Knowledge," *Environment and Planning D: Society and Space* 10, no.1 (1992): 23–40.

104. Trevor J. Barnes and Matthew Farish, "Between Regions: Science, Militarism, and American Geography from World War to Cold War," *Annals of the Association of American Geographers* 96, no. 4 (2006): 808.

105. Judith Butler, *Frames of War: When Is Life Grievable?* (London: Verso, 2009), xiii.

106. Derek Gregory, "The Everywhere War," *Geographical Journal* 177, no. 3 (2011): 238–50.

107. Henri Lefebvre, *The Production of Space,* trans. Donald Nicholson-Smith (Oxford: Blackwell, 1991), 196.

108. Sloterdijk, *In the World Interior of Capital,* 13.

109. Ibid., 46.

110. Ibid., 31.

111. Michel Serres, *The Parasite,* trans. Lawrence R. Schehr (Minneapolis: University of Minnesota Press, 2007), 59.

112. Michael S. Reidy, *Tides of History: Ocean Science and Her Majesty's Navy* (Chicago: University of Chicago Press, 2009), 6.

113. Ibid., 6.

114. Sloterdijk, *In the World Interior of Capital,* 110.

115. Ibid., 140.

116. Martin Heidegger, *The Question Concerning Technology and Other Essays,* trans. William Lovitt (New York: Garland Publishing, 1977), 134.

117. Simon Dalby, "Political Space," 433.

118. Stuart Elden, *Terror and Territory: The Spatial Extent of Sovereignty* (Minneapolis: University of Minnesota Press, 2009), xxvii.

119. Dalby, "Political Space ," 436.

120. Stuart Elden, "Missing the Point: Globalization, Deterritorialization, and the Space of the World," *Transactions of the Institute of British Geographers* 30, no. 1 (2005): 8–19.

121. Ibid., 16.

122. Sloterdijk, *In the World Interior of Capital,* 30.

123. Rey Chow, *The Age of the World Target: Self-Referentiality in War, Theory, and Comparative Work* (Durham, N.C.: Duke University Press, 2006).

124. Gilbert Simondon, *On the Mode of Existence of Technical Objects,* trans. Ninian Mellamphy (London, Ontario: University of Western Ontario, 1980), 4.

125. Simondon, *On the Mode of Existence of Technical Objects,* 18.

126. Amy E. Wendling, *Karl Marx on Technology and Alienation* (Basingstoke, U.K.: Palgrave Macmillan, 2009), 56.

127. Ibid., 140.

128. Wolfgang Schivelbusch, *The Railway Journey: The Industrialization of Time and Space in the 19th Century* (Berkeley: University of California Press, 1986), 40.

129. Wendling, *Karl Marx on Technology and Alienation,* 193.

130. William E. Connolly, "'The New Materialism' and the Fragility of Things," *Millennium: Journal of International Studies* 41, no. 3 (2013): 403.

131. Elizabeth Grosz, "Identity and Individuation," in *Gilbert Simondon: Being and Technology,* ed. Arne De Boever et al. (Edinburgh: Edinburgh University Press, 2012), 43.

132. Ibid., 50.

133. Muriel Combes, *Gilbert Simondon and the Philosophy of the Transindividual,* trans. Thomas LaMarre (Cambridge, Mass.: MIT Press, 2013), 9.
134. Foucault, *"Society Must Be Defended,"* 29.
135. Nigel Clark, *Inhuman Nature: Sociable Life on a Dynamic Planet* (London: Sage, 2011), 27.
136. Bernard Stiegler, *Technics and Time,* vol. 1, *The Fault of Epimetheus,* trans. Richard Beardsworth and George Collins (Stanford, Calif.: Stanford University Press, 1998), 177.
137. Gilles Deleuze, *Negotiations* (New York: Columbia University Press, 1995), 117.
138. Guattari, *The Three Ecologies,* 39.
139. Ibid., 50.
140. Agamben, *What Is an Apparatus?,* 14.
141. Ibid., 15.
142. Ibid., 17.
143. Ibid., 17.
144. Ibid., 20.
145. Žižek, *Violence,* 2.
146. Ellul, *The Technological Society,* 102.
147. Ibid., 100.
148. Ibid., 100.
149. Guattari, *The Three Ecologies,* 42.
150. Ellul, *The Technological Society,* 428.
151. Freud, *Civilization and Its Discontents,* 87.
152. Ibid., 105.
153. Roberto Esposito, *Bíos: Biopolitics and Philosophy,* trans. Timothy Campbell (Minneapolis: University of Minnesota Press, 2008), 55.

2. The Rise of the Predator Empire in the Vietnam War

1. James W. Gibson, *The Perfect War: Technowar in Vietnam,* 2nd ed. (New York: Atlantic Monthly Press, 2000).
2. Paul Dickson, *The Electronic Battlefield* (Takoma Park, Md.: FoxAcre Press, 2012), 17.
3. William C. Westmoreland, address at the Association of the United States Army Annual Luncheon, Sheraton Park Hotel, Washington, D.C., October 14, 1969, quoted in Dickson, *The Electronic Battlefield,* 220–21.
4. Karl Marlantes, *Matterhorn* (New York: Atlantic Monthly Press, 2010), 343.
5. Andrew Wiest, introduction to *Rolling Thunder in a Gentle Land: The Vietnam War Revisited,* ed. Andrew Wiest (Oxford: Osprey Publishing, 2006), 25.
6. Gibson, *The Perfect War,* 63.
7. Bui Tin, "Fight for the Long Haul," in *Rolling Thunder in a Gentle Land,* 53.

8. Le Ly Hayslip and Dien Pham, "Caught in the Crossfire," in *Rolling Thunder in a Gentle Land,* 138.

9. Wiest, introduction to *Rolling Thunder,* 29.

10. Andrew J. Bacevich, *Washington Rules: America's Path to Permanent War* (New York: Metropolitan Books, 2010), 138.

11. Wiest, introduction to *Rolling Thunder,* 30.

12. Nick Turse, *Kill Anything That Moves: The Real American War in Vietnam* (New York: Metropolitan Books, 2013), 10–11.

13. Gibson, *The Perfect War,* vii.

14. Ibid., 23.

15. Gordon L. Rottman, "Tactics in a Different War," in *Rolling Thunder in a Gentle Land,* 230–31.

16. International Committee of the Red Cross, *Weapons That May Cause Unnecessary Suffering or Have Indiscriminate Effects* (Geneva, Switzerland: International Committee of the Red Cross, 1973), 60.

17. Ibid., 61.

18. Antoine Bousquet, *The Scientific Way of Warfare: Order and Chaos on the Battlefields of Modernity* (London: Hurst, 2009), 149.

19. Turse, *Kill Anything That Moves,* 42.

20. Bousquet, *The Scientific Way of Warfare,* 149.

21. Alfred W. McCoy, "Imperial Illusions: Information Infrastructure and the Future of U.S. Global Power," in *Endless Empire: Spain's Retreat, Europe's Eclipse, America's Decline,* ed. Alfred W. McCoy, Josep M. Fradera, and Stephen Jacobson (Madison: University of Wisconsin Press, 2012), 370.

22. Bousquet, *The Scientific Way of Warfare,* 161.

23. Turse, *Kill Anything That Moves,* 45.

24. Ibid., 209.

25. Turse, *Kill Anything That Moves,* 80.

26. Derek Gregory, "Lines of Descent," *OpenDemocracy,* November 8, 2011, https://www.opendemocracy.net.

27. Matthew A. Kocher, Thomas B. Pepinsky, and Stathis N. Kalyvas, "Aerial Bombing and Counterinsurgency in the Vietnam War," *American Journal of Political Science* 55, no. 2 (2011): 201–18.

28. Arthur H. Westing, "The Environmental Aftermath of Warfare in Viet Nam," *Natural Resources Journal* 23 (1983): 365–89.

29. Ronald B. Frankum, "'Swatting Flies with a Sledgehammer,'" in *Rolling Thunder in a Gentle Land,* 193.

30. Derek Gregory, "Lines of Descent."

31. Turse, *Kill Anything That Moves,* 62.

32. Bernard C. Nalty, *The War against Trucks: Aerial Interdiction in Southern Laos, 1968–1972* (Washington, D.C.: Air Force Historical Studies Office, 2005), 47–51.

33. Frankum, "Swatting Flies with a Sledgehammer," 200.

34. Chalmers Johnson, *Blowback: The Costs and Consequences of American Empire,* 2nd ed. (London: Time Warner Paperbacks, 2002), 12.

35. John W. Finney, "Guided Bombs Expected to Revolutionize Warfare," *New York Times,* March 18, 1974.

36. "Army Research Yields Laser-Guided Missile Technology," *Army Research and Development Magazine,* September 1972, 5.

37. DMS Market Intelligence Report, *Electronic Systems,* February 1972.

38. Glenn Rutherford, "Army Developing High Energy Lasers," *Louisville Courier Journal,* November 18, 1974.

39. Orr Kelly, "Even Smarter 'Smart Bombs,'" *Washington Evening Star,* October 3, 1972.

40. Robert Hotz, "New Lessons from Vietnam," *Aviation Week and Space Technology,* May 22, 1972, 7.

41. Gibson, *The Perfect War,* 123.

42. Ibid., 225.

43. Turse, *Kill Anything That Moves,* 93.

44. Yves Lacoste, "An Illustration of Geographical Warfare: Bombing of the Dikes on the Red River, North Vietnam," *Antipode* 5, no. 2 (1973): 2.

45. Peter Sloterdijk, *Terror from the Air,* trans. Amy Patton and Steve Corcoran (Los Angeles: Semiotext[e], 2009), 14.

46. Michael G. Palmer, "The Legacy of Agent Orange: Empirical Evidence from Central Vietnam," *Social Science and Medicine* 60 (2005): 1061.

47. Turse, *Kill Anything That Moves,* 96–97.

48. Ibid., 83, 96.

49. Arthur H. Westing, "Environmental Consequences of the Second Indochina War: A Case Study," *Ambio* 4 (1975): 221–22.

50. Westing, *The Environmental Aftermath of Warfare in Viet Nam,* 375.

51. Nalty, *The War against Trucks,* 33.

52. "Pentagon Confirms Tear Gas Bombing," *Washington Star,* July 14, 1972.

53. D. Hank Ellison, *Chemical Warfare during the Vietnam War: Riot Control Agents in Combat* (New York: Routledge, 2012), 50.

54. Edmund Stillman, "Smart Bombs and Dumb Strategy," *Saturday Review of the Society,* July 29, 1972, 30.

55. Dickson, *The Electronic Battlefield,* 31.

56. Jacob Van Staaveren, *Interdiction in Southern Laos, 1960–1968: The United States Air Force in Southeast Asia,* Center for Air Force History (Washington, D.C.: U.S. Government Printing Office, 1993), 264.

57. Dickson, *The Electronic Battlefield,* 83.

58. Van Staaveren, *Interdiction in Southern Laos,* 259.

59. *Investigation into Electronic Battlefield Program: Hearings before the Electronic Battlefield Subcommittee of the Preparedness Investigating Subcommittee of the Committee on Armed Services, United States Senate, Ninety-Second Congress* (Washington, D.C.: U.S. Government Printing Office, 1971), 11.

60. Colonel McCoskrie, quoted in Nalty, *The War against Trucks,* 23.

61. Van Staaveren, *Interdiction in Southern Laos,* 271.

62. Ibid., 271.

63. Nalty, *The War against Trucks,* 123.

64. Dickson, *The Electronic Battlefield,* 85.

65. Gibson, *The Perfect War,* 397.

66. General Slay, quoted in Nalty, *The War against Trucks,* 294.

67. Robert Barkan, “Bringing the Toys Home,” *Pacific Research and World Empire Telegram,* November–December 1971.

68. Ken W. Clawson, “U.S. Testing Sensors along Mexican Border,” *Washington Post,* July 18, 1970.

69. Raul Ramirez, “Washable Bullet-Proof Vest Comes in Light Blue for $99,” *Washington Post,* October 3, 1972.

70. “Police Advances Bring a Warning; Engineer Fears Technology Could Lead to Oppression,” *New York Times,* November 11, 1967.

71. “Night Vision Devices,” *Army Research and Development News Magazine,* July 1972, 10.

72. Mike Clark, “Electronics Moves from War in Vietnam,” *Philadelphia Inquirer,* February 20, 1972.

73. Joseph A. Meyer, “Crime Deterrent Transponder System,” *IEEE Transactions of Aerospace and Electronic Systems* 7, no. 1 (1971): 2.

74. Ibid., 5.

75. Ibid., 17.

76. *Aerospace Daily and Defense Report,* July 20, 1972, 97.

77. Phil Stanford, “The Automated Battlefield,” *New York Times,* February 23, 1975.

78. Seymour J. Deitchman, “The ‘Electronic Battlefield’ in the Vietnam War,” *Journal of Military History* 72, no. 3 (2008): 869–87.

79. William E. DePuy to Gen. Creighton W. Abrams, January 14, 1974, quoted in Robert R. Tomes, “An Historical Review of US Defense Strategy from Vietnam to Operation Iraqi Freedom,” *Defense and Security Analysis* 28, no. 4 (2012): 304.

80. Stanford, “The Automated Battlefield.”

81. Ibid.

82. Ibid.

83. “Visions of the Next War,” *Newsweek,* April 22, 1974, 53.

84. *Aviation Week and Space Technology,* July 29, 1974, 57.

85. Barry D. Watts, *The Maturing Revolution in Military Affairs* (Washington, D.C.: Center for Strategic and Budgetary Assessments, 2011), 1–2, www.csbaonline.org..

86. Douglas McGray, “The Marshall Plan,” *Wired,* February 2003, archive .wired.com.

87. Keith L. Shimko, *The Iraq Wars and America’s Military Revolution* (Cambridge: Cambridge University Press, 2010), 97.

88. Thomas A. Keaney and Eliot A. Cohen, *Gulf War Air Power Survey Summary Report,* Air Force Historical Studies Office (Washington, D.C.: U.S. Government Printing Office, 1993), 248.

89. Ibid., 241.

90. Ibid., 226.

91. Watts, *The Maturing Revolution in Military Affairs,* 32.

92. William A. Owens, "The Emerging U.S. System-of-Systems," *Strategic Forum* 63 (1996): 1–2.

93. Arthur K. Cebrowski and John H. Garstka, "Network-Centric Warfare: Its Origin and Future," *Proceedings Magazine,* January 1998, www.usni.org/.

94. Peter Warren Singer, *Wired for War: The Robotics Revolution and Conflict in the 21st Century* (New York: Penguin, 2009), 180.

95. John Arquilla and David Ronfeldt, *Networks and Netwars: The Future of Terror, Crime, and Militancy* (Santa Monica, Calif.: RAND Corporation, 2001), 1.

96. George W. Bush, "Bush Addresses the Citadel" (speech, Charleston, S.C., December 11, 2001), CNN.com/Transcripts, http://edition.cnn.com/TRANSCRIPTS/0112/11/se.04.html.

97. Shimko, *The Iraq Wars and America's Military Revolution,* 234.

98. Ibid., 163.

99. Ibid., 164.

100. Bacevich, *Washington Rules,* 181.

101. Valentine, *The Phoenix Program,* 420.

102. Col. Andrew R. Finlayson, "A Retrospective on Counterinsurgency Operations: The Tay Ninh Provincial Reconnaissance Unit and Its Role in the Phoenix Program, 1969–70," *Studies of Intelligence* 51, no. 2 (2007): www.cia.gov.

103. Stuart A. Herrington, *Silence Was a Weapon: The Vietnam War in the Villages* (New York: Ballantine Books, 1982), 53.

104. Frances Fitzgerald, *Fire in the lake: The Vietnamese and the Americans in Vietnam* (New York: Back Bay Books / Little, Brown, 1972), 412.

105. Valentine, *The Phoenix Program,* 34.

106. Seymour M. Hersh, "Moving Targets," *New Yorker,* December 15, 2003, www.newyorker.com.

107. Valentine, *The Phoenix Program,* 81.

108. Ibid., 85.

109. Finlayson, *A Retrospective on Counterinsurgency Operations.*

110. Alfred W. McCoy, "Confronting the CIA's Mind Maze," *TomDispatch,* June 7, 2009, www.tomdispatch.com.

111. Jane Mayer, "The Black Sites," *New Yorker,* August 13, 2007, www.newyorker.com.

112. McCoy, "Imperial Illusions," 362.

113. Ibid., 368.

114. Valentine, *The Phoenix Program,* 259.

115. Dana Priest, "Army's Project X Had Wider Audience," *Washington Post,* March 6, 1997.

116. Alfred W. McCoy, *Torture and Impunity: The U.S. Doctrine of Coercive Interrogation* (Madison: University of Wisconsin Press, 2012), 100.

117. Mary Turck, "School of Assassins," *Common Dreams,* November 18, 2003.

118. McCoy, *Torture and Impunity*, 103.

119. Michael Hastings, "The Rise of Killer Drones: How America Goes to War in Secret," *Rolling Stone,* April 16, 2012, www.rollingstone.com.

120. "Kettering Aerial Torpedo 'Bug' Fact Sheet," National Museum of the U.S. Air Force website, March 25, 2014, www.nationalmuseum.af.mil.

121. Dickson, *The Electronic Battlefield,* 181.

122. Thomas P. Ehrhard, *Air Force UAVs: The Secret History* (Arlington, Va.: Mitchell Institute Press, 2010).

123. "U.S. Is Flying Pilotless Planes over Viet Nam," *Washington Daily News,* November 18, 1964.

124. "The Firebee," *Time,* November 27, 1964.

125. Comptroller General of the United States, *DOD's Use of Remotely Piloted Vehicle Technology Offers Opportunities for Saving Lives and Dollars,* MASAD-81-20, April 3, 1981, 2, http://archive.gao.gov/f0102/114831.pdf.

126. Dickson, *The Electronic Battlefield,* 188.

127. Ehrhard, *Air Force UAVs,* 23.

128. Scott Orr, "DRONES DECLASSIFIED: UAVs date back to '60s over China, Korea, Vietnam," *Daily Courier,* January 17, 2014, www.dcourier.com.

129. Barry Miller, "USAF Widens Unmanned Aircraft Effort," *Aviation Week and Space Technology,* November 9, 1970.

130. Ehrhard, *Air Force UAVs,* 24.

131. Ibid., 28.

132. Brian Fung, "The CIA First Tested Drones in Area 51 Because of Course They Did," *Washington Post,* August 16, 2013, www.washingtonpost.com.

133. Michael Getler, "Air Force Eyes 'Drone' Craft," *Washington Post,* May 6, 1974.

134. William A. Anderson, "Age of the RPV: A Special Report," *Teledyne Ryan Aeronautical Reporter* 32 (1971): 2.

135. *U.S. News and World Report,* February 28, 1972, 57.

136. Edgar Ulsamer, "Status Report on Laser Weapons," *Air Force Magazine,* January 1972, 40.

137. Tim Blackmore, *War X: Human Extensions in Battlespace* (Toronto: University of Toronto Press, 2005) 148–49.

138. Comptroller General of the United States, *DOD's Use of Remotely Piloted Vehicle Technology Offers Opportunities for Saving Lives and Dollars.*

139. See "The Dronefather," *Economist,* December 1, 2012, www.economist.com; Peter Finn, "Rise of the Drone: From Calif. Garage to Multibillion-Dollar Defense Industry," *Washington Post,* December 23, 2011, www.washingtonpost.com.

140. "The Dronefather."

141. Ibid.

142. Mark Mazzetti, *The Way of the Knife: The CIA, a Secret Army, and a War at the Ends of the Earth* (New York: Penguin, 2013), 91.

143. Finn, "Rise of the Drone."

144. Anthony H. Cordesman, *The Lessons and Non-lessons of the Air and Missile Campaign in Kosovo.* Center for Strategic and International Studies website, September 17, 2003, 321, http://csis.org/publication/lessons-and-non-lessons-air-and-missile-campaign-kosovo.

145. Ibid.

146. "The Dronefather."

147. Mazzetti, *The Way of the Knife,* 100.

148. Singer, *Wired for War,* 194.

149. Dickson, *The Electronic Battlefield*, 208–9.

3. Full Spectrum Global Dominance

1. Philip Alston, *Report of the Special Rapporteur on Extrajudicial, Summary, or Arbitrary Executions, Addendum: Study on Targeted Killings,* United Nations General Assembly, Human Rights Council, 14th session, Agenda item 3, GE.10–13753, May 28, 2010, 3.

2. Gabriella Blum, "The Individualization of War: From War to Policing in the Regulation of Armed Conflict," in *Law and War,* ed. Austin Sarat, Lawrence Douglas, and Martha Merrill Umphrey (Stanford, Calif.: Stanford University Press, 2014), 48.

3. Ibid., 49.

4. Alston, *Report of the Special Rapporteur,* 3.

5. Micah Zenko, *Reforming U.S. Drone Strike Policies,* Council on Foreign Relations Special Report No. 65 (2013), 8.

6. Jonathan Masters, "Targeted Killings," Council on Foreign Relations, May 23, 2013, www.cfr.org.

7. Eyal Weizman, *Hollow Land: Israel's Architecture of Occupation* (London: Verso, 2007), 241.

8. George A. Crawford, *Manhunting: Counter-Network Organization for Irregular Warfare,* JSOU Report 09-7 (Hurlburt Field, Fla.: JSOU Press 2009), 1.

9. Grégoire Chamyou, *Manhunts: A Philosophical History,* trans. Steven Rendall (Princeton, N.J.: Princeton University Press, 2012), 27.

10. Grégoire Chamyou, *Drone Theory,* trans. Janet Loyd (New York: Penguin, 2015), 53.

11. Ibid., 52.

12. A transcript of *A Study of Assassination* is available through the National Security Archive at www2.gwu.edu/~nsarchiv/NSAEBB/NSAEBB4/ciaguat2.html.

13. Seymour M. Hersh, "Huge CIA Operation Reported in U.S. against Antiwar Forces, Other Dissidents in Nixon Years," *New York Times,* December 22, 1974.

14. President Ronald Reagan, "Message on the Observance of Afghanistan Day" (speech, Washington D.C., March 21, 1983), Ronald Reagan Presidential Library, www.reagan.utexas.edu/.

15. Steve Coll, *Ghost Wars: The Secret History of the CIA, Afghanistan, and Bin Laden, from the Soviet Invasion to September 10, 2001* (London: Penguin Books, 2004), 145.

16. President Ronald Reagan, *The National Program for Combating Terrorism,* National Security Decision Directive Number 207, Washington, D.C., January 20, 1986, Ronald Reagan Presidential Library, www.reagan.utexas.edu/archives/reference/Scanned%20NSDDS/NSDD207.pdf.

17. Mark Mazzetti, *The Way of the Knife: The CIA, a Secret Army, and a War at the Ends of the Earth* (New York: Penguin, 2013), 57.

18. Coll, *Ghost Wars,* 423–27.

19. Jane Mayer, "The Black Sites," *New Yorker,* August 13, 2007, www.newyorker.com.

20. Bob Woodward, "CIA Led Way with Case Handouts," *Washington Post,* November 18, 2002, www.washingtonpost.com.

21. George W. Bush, "Bush Addresses the Citadel," (speech, Charleston, S.C., December 11, 2001), CNN.com/Transcripts, http://edition.cnn.com/TRANSCRIPTS/0112/11/se.04.html.

22. John Sifton, "A Brief History of Drones," *Nation,* February 7, 2012, www.the nation.com.

23. James Risen and David Johnston, "Bush Has Widened Authority of C.I.A. to Kill Terrorists," *New York Times,* December 15, 2002, www.nytimes.com.

24. Seymour M. Hersh, "Manhunt," *New Yorker,* December 23, 2002, www.newyorker.com.

25. Donald H. Rumsfeld, "Secretary Rumsfeld's Media Roundtable at Camp Victory, Iraq" (roundtable, Camp Victory, Iraq, September 13, 2002), U.S. Department of Defense, www.defense.gov/transcripts/.

26. Mayer, "The Black Sites."

27. Hakim Almasmari, Margaret Coker, and Siobhan Gorman, "Drone Kills Top Al Qaeda Figure," *Wall Street Journal,* October 1, 2011, www.online.wsj.com.

28. Condoleeza Rice, interview by Tony Snow, *Fox News Sunday,* November 10, 2002, www.foxnews.com.

29. Andrew Bacevich, "Scoring the 'War on Terror,'" *Asia Times,* February 25, 2012, www.atimes.com.

30. Mazzetti, *The Way of the Knife,* 19.

31. Seymour M. Hersh, "The Gray Zone," *New Yorker,* May 24, 2004, www.newyorker.com.

32. Seymour M. Hersh, "The General's Report," *New Yorker,* June 25, 2007, www.newyorker.com.

33. Barton Gellman, "Secret Unit Expands Rumsfeld's Domain," *Washington Post,* January 23, 2005, www.washingtonpost.com.

34. Jeremy Scahill, *Dirty Wars: The World Is a Battlefield* (London: Serpent's Tail, 2013), 101.

35. Nick Turse, "The Golden Age of Black Ops," *TomDispatch,* January 20, 2015, www.tomdispatch.com.

36. Dana Priest and William M. Arkin, "'Top Secret America': A Look at the Military's Joint Special Operations Command," *Washington Post,* September 2, 2011, www.washingtonpost.com.

37. Eric Schmitt and Mark Mazzetti, "Secret Order Lets US Raid Al Qaeda," *New York Times,* November 9, 2008, www.nytimes.com.

38. Scahill, *Dirty Wars,* 172.

39. Mazzetti, *The Way of the Knife,* 74–75.

40. Gellman, "Secret Unit Expands Rumsfeld's Domain."

41. Priest and Arkin, "Top Secret America."

42. Mazzetti, *The Way of the Knife,* 77.

43. Priest and Arkin, "Top Secret America."

44. Scahill, *Dirty Wars,* 168.

45. Seymour M. Hersh, "Moving Targets," *New Yorker,* December 15, 2003, www.newyorker.com.

46. Douglas Valentine, "Preemptive Manhunting, the CIA's New Assassination Program," *CounterPunch,* December 11, 2003, www.counterpunch.org.

47. Julian Coman, "CIA Plans New Secret Police to Fight Iraq Terrorism," *Telegraph,* January 4, 2004, www.telegraph.co.uk.

48. Scahill, *Dirty Wars,* 151.

49. Nick Davies, "Afghanistan War Logs: Task Force 373—Special Forces Hunting Top Taliban," *Guardian,* July 25, 2010, www.theguardian.com.

50. Scahill, *Dirty Wars,* 331.

51. Gretchen Gavett, "What Is the Secretive U.S. 'Kill/Capture' Campaign?," *Frontline,* June 17, 2011, www.pbs.org.

52. Masters, "Targeted Killings."

53. Jim Michaels, "New Training Hones Marines' Visual Skills," *USA Today,* June 26, 2008, www.usatoday.com.

54. Ian Cobain, "CIA Rendition: More Than a Quarter of Countries 'Offered Covert Support'," *Guardian,* February 5, 2013.

55. Mazzetti, *The Way of the Knife,* 121.

56. Greg Miller and Julie Tate, "CIA Shifts Focus to Killing Individuals," *Washington Post,* September 1, 2011, www.washingtonpost.com.

57. Mazzetti, *The Way of the Knife,* 4.

58. Scahill, *Dirty Wars,* 516–517.

59. Chamayou, *Drone Theory,* 52.

60. Ibid., 35.

61. Causality figures and data on drone strikes in Pakistan, Yemen, Somalia, and Afghanistan in this chapter come from the *Bureau of Investigative Journalism*'s regularly updated database, http://www.thebureauinvestigates.com/category/projects/drones/.

62. Ian G. R. Shaw, and Majed Akhter, "The Unbearable Humanness of Drone Warfare in FATA, Pakistan," *Antipode* 44, no. 4 (2012): 1490–509.

63. Priya Satia, "Drones: A History from the British Middle East," *Humanity: An International Journal of Human Rights, Humanitarianism, and Development* 5, no. 1 (2014): 2.

64. R. H. Peck, "Aircraft in Small Wars." *Royal United Services Institution Journal* 73, no. 491 (1928): 537.

65. Graham Chandler, "The Bombing of Waziristan," *Air and Space Magazine,* July 2011, www.airspacemag.com.

66. Satia, "Drones," 6.

67. Ian G. R. Shaw and Majed Akhter, "The Dronification of State Violence," *Critical Asian Studies* 46, no. 2 (2014): 211–34.

68. Miller and Tate, "CIA Shifts Focus to Killing Targets."

69. International Human Rights and Conflict Resolution Clinic at Stanford Law School and Global Justice Clinic at NYU School of Law, *Living under Drones: Death, Injury, and Trauma to Civilians from U.S. Drone Practices in Pakistan* (Stanford, Calif: International Human Rights and Conflict Resolution Clinic at Stanford Law School and Global Justice Clinic at NYU School of Law, 2012), 98.

70. Satia, "Drones," 20.

71. Mark Neocleous, *War Power, Police Power* (Edinburgh: Edinburgh University Press, 2014), 162.

72. Alfred W. McCoy, *Policing America's Empire: The United States, the Philippines, and the Rise of the Surveillance State* (Madison: University of Wisconsin Press, 2009), 19.

73. Michel Foucault, *"Society Must Be Defended": Lectures at the Collège de France, 1975–1976,* trans. David Macey (London: Penguin, 2003), 244.

74. Louise Amoore, "Algorithmic War: Everyday Geographies of the War on Terror," *Antipode* 41, no. 1 (2009): 49–69.

75. "About NGA," National Geospatial-Intelligence Agency, www.nga.mil.

76. Linda D. Kozaryn, "Force XXI Commander Heads New Age Brigade," American Forces Press Service, U.S. Department of Defense, March 27, 1997, www.defense.gov.

77. Jeremy W. Crampton, Sue M. Roberts, and Ate Poorthuis, "The New Political Economy of Geographical Intelligence," *Annals of the Association of American Geographers* 104, no. 1 (2014): 206.

78. Derek Gregory, "From a View to a Kill: Drones and Late Modern War," *Theory, Culture, and Society* 28, nos. 7–8 (2011): 195.

79. Ian G. R. Shaw, "Predator Empire: The Geopolitics of U.S. Drone War," *Geopolitics* 18, no. 3 (2013): 536–39.

80. Michael Dillon, "Governing Terror: The State of Emergency of Biopolitical Emergence," *International Political Sociology* 1, no. 1 (2007): 24.

81. Michael Hardt and Antonio Negri, *Empire* (Cambridge, Mass.: Harvard University Press, 2000), 58.

82. Chalmers Johnson, *The Sorrows of Empire: Militarism, Secrecy, and the End of the Republic* (London: Verso, 2004), 23.

83. Chalmers Johnson, *Blowback: The Costs and Consequences of American Empire,* 2nd ed. (London: Time Warner Paperbacks, 2002), 37.

84. Ibid., xvii.

85. John Morrissey, "Liberal Lawfare and Biopolitics: US Juridical Warfare in the War on Terror," *Geopolitics* 16 (2011): 287.

86. U.S. Department of Defense, *Base Structure Report—Fiscal Year 2014 Baseline.*

87. Johnson, *The Sorrows of Empire,* 188.

88. Nick Turse, "Empire of Bases 2.0," *TomDispatch,* January 9, 2011, www.tomdispatch.com.

89. David Vine, "Garrisoning the Globe," *TomDispatch,* September 13, 2015, www.tomdispatch.com.

90. Dan Froomkin, "U.S. to Hand Over Iraq Bases, Equipment Worth Billions," *Huffington Post,* November 26, 2011, www.huffingtonpost.com.

91. Eric Schmitt and Carolyn Marshall, "In Secret Unit's 'Black Room,' a Grim Portrait of U.S. Abuse," *New York Times,* March 19, 2006, www.nytimes.com.

92. Seymour M. Hersh, "Torture at Abu Ghraib," *New Yorker,* May 10, 2004, www.newyorker.com.

93. Martin Chulov, "Isis: The Inside Story," *Guardian,* December 11, 2014, www.theguardian.com.

94. Thomas E. Ricks, "Biggest Base in Iraq Has Small-Town Feel," *Washington Post,* February 4, 2006, www.washingtonpost.com.

95. Mark Urban, *Task Force Black: The Explosive True Story of the SAS and the Secret War in Iraq* (London: Little, Brown, 2010), 82, quoted in Scahill, *Dirty Wars,* 162.

96. Nick Turse, "Afghanistan's Base Bonanza," *TomDispatch,* September 4, 2012, www.tomdispatch.com.

97. Ibid.

98. Scahill, *Dirty Wars,* 331.

99. Amy Kaplan, "Where Is Guantánamo?," *American Quarterly* 57 (2005): 837.

100. Ibid., 844.

101. Tom Engelhardt, "The Great American Disconnect," *TomDispatch,* June 7, 2007, www.tomdispatch.com.

102. Morrissey, "Liberal Lawfare and Biopolitics," 291.

103. Chalmers Johnson, "America's Empire of Bases," *TomDispatch,* January 15, 2004, www.tomdispatch.com.

104. Nick Turse, "America's Secret Empire of Drone Bases," *TomDispatch,* October 16, 2011, www.tomdispatch.com.

105. Nick Turse, "450 Bases and It's Not Over Yet," *TomDispatch,* February 12, 2012, www.tomdispatch.com.

106. Chris Woods and Alice K. Ross, "Revealed: U.S. and Britain Launched 1,200 Drone Strikes in Recent Wars," *Bureau of Investigative Journalism,* December 4, 2012, www.thebureauinvestigates.com. Statistics on drone strikes are subject to regular revision on the website of the *Bureau of Investigative Journalism.*

107. Ibid.

108. "President Obama Speaks with VICE News," interviewed by Shane Smith, Vice News, March 17, 2015, https://news.vice.com.

109. Up-to-date figures and statistics are currently maintained at http://airwars.org.

110. Craig Whitlock, "U.S. Relies on Persian Gulf Bases for Airstrikes in Iraq," *Washington Post,* August 26, 2014, www.washingtonpost.com.

111. Pratap Chatterjee, "How Lawyers Sign Off on Drone Attacks," *Guardian,* June 15, 2011, www.theguardian.com.

112. Chris Woods, "CIA's Pakistan Drone Strikes Carried Out by Regular US Air Force Personnel," *Guardian,* April 14, 2014, www.theguardian.com.

113. U.S. Air Force, "Air Force Distributed Common Ground System," www.af.mil.

114. Barack Obama, "Presidential Letter—2012 War Powers Resolution 6-Month Report," White House, Office of the Press Secretary, June 15, 2012, www.whitehouse.gov.

115. Human Rights Watch, *"Between a Drone and Al-Qaeda": The Civilian Cost of US Targeted Killings in Yemen,* October 22, 2013, https://www.hrw.org/report/2013/10/22/between-drone-and-al-qaeda/civilian-cost-us-targeted-killings-yemen.

116. "Yemen: Reported US Covert Actions 2001–2011," *Bureau of Investigative Journalism,* March 29, 2012, www.thebureauinvestigates.com.

117. Eric H. Holder to Patrick J. Leahy, May 22, 2013.

118. Greg Miller, "White House Approves Broader Yemen Drone Campaign," *Washington Post,* April 26, 2012, www.washingtonpost.com.

119. Human Rights Watch, *A Wedding That Became a Funeral: US Drone Attack on Marriage Procession in Yemen,* February 19, 2014, https://www.hrw.org/report/2014/02/19/wedding-became-funeral/us-drone-attack-marriage-procession-yemen.

120. "Covert Drone War," *Bureau of Investigative Journalism* website, updated regularly, http://www.thebureauinvestigates.com/category/projects/drones/.

121. David Alexander, "Retired General Cautions against Overuse of 'Hated' Drones," Reuters, January 7, 2013, www.reuters.com.

122. Rebecca Kaplan, "Obama: U.S. Can't 'Play Whack-a-Mole' with Militant Groups," *CBS News,* June 22, 2014, www.cbsnews.com.

123. Jeff Faux, "Whack-A-Mole: Our Strategy for Permanent War," *Huffington Post,* November 29, 2014, www.huffingtonpost.com.

124. Chamayou, *Drone Theory,* 71.

125. Whitlock, "U.S. Expands Secret Intelligence Operations in Africa."

126. Nick Turse, "The Secret U.S. Military Operation Underway in Africa," *Mother Jones,* May 15, 2014, www.motherjones.com.

127. Nick Turse, *The Changing Face of Empire: Special Ops, Drones, Spies, Proxy Fighters, Secret Bases, and Cyberwarfare* (Chicago: Haymarket Books, 2012), 68.

128. Nick Turse, "2044 or Bust," *TomDispatch,* April 14, 2015, www.tomdispatch.com.

129. Craig Whitlock, "Remote U.S. Base at Core of Secret Operations," *Washington Post,* October 26, 2012, www.washingtonpost.com.

130. David S. Cloud, "U.S. Military Presence in Africa Growing in Small Ways," *Los Angeles Times,* March 7, 2014, www.latimes.com.

131. Whitlock, "Remote U.S. Base at Core of Secret Operations."

132. Nick Turse, "America's Secret African Drone War Against the Islamic State," *TomDispatch,* December 17, 2015, www.tomdispatch.com.

133. Sean D. Naylor, "Clandestine Somalia Missions Yield AQ Targets," *Army Times,* November 14, 2011, www.armytimes.com.

134. Jeremy Scahill, "The CIA's Secret Sites in Somalia," *Nation,* December 10, 2014, www.thenation.com.

135. Scahill, *Dirty Wars,* 124.

136. Ibid., 208.

137. Michael R. Gordon and Mark Mazzetti, "U.S. Used Base in Ethiopia to Hunt Al Qaeda," *New York Times,* February 23, 2007, www.nytimes.com.

138. Scahill, *Dirty Wars,* 494.

139. "Covert Drone War," *Bureau of Investigative Journalism* website.

140. Jack Serle, "Does Latest Drone Strike on al Shabaab Signal Change in U.S. Tactics in Somalia?," *Bureau of Investigative Journalism,* February 6, 2015, www.thebureauinvestigates.com.

141. Craig Whitlock, "U.S. Expands Secret Intelligence Operations in Africa," *Washington Post,* June 13, 2012, www.washingtonpost.com.

142. Nick Turse, "Washington's Back-to-the-Future Military Policies in Africa," *TomDispatch,* March 13, 2014, www.tomdispatch.com.

143. Karen DeYoung, "Obama Boosts Effort to Find Kony," *Washington Post,* March 23, 2014, www.washingtonpost.com.

144. Craig Whitlock, "Contractors Run U.S. Spying Missions in Africa," *Washington Post,* June 14, 2012, www.washingtonpost.com.

145. Nick Turse, "AFRICOM Goes to War on the Sly," *TomDispatch,* April 13, 2014, www.tomdispatch.com.

146. Mark Mazzetti and Eric Schmitt, "Pentagon Seeks to Knit Foreign Bases into ISIS-Foiling Network," *New York Times,* December 10, 2015, www.nytimes.com.

147. President George Washington to Marquis de Lafayette, November 15, 1781, *The Writings of George Washington from the Original Manuscript Sources 1745–1799,* vol. 23 (Washington, DC: Government Printing Office, 1937), 341.

148. "Fact File: Aircraft Carriers—CVN," U.S. Navy website, October 16, 2014, www.navy.mil.

149. "Why the Carriers?," U.S. Navy, accessed October 16, 2014, http://www.navy.mil/navydata/ships/carriers/cv-why.asp.

150. Ed Pilkington, "Obama under Fire over Detention of Terror Suspect on US Navy Ship," *Guardian,* July 6, 2011, www.theguardian.com.

151. Tom Engelhardt, "Offshore Everywhere," *TomDispatch,* February 5, 2012, www.tomdispatch.com.

152. U.S. Department of Defense, "Contracts," no. 582-13, August 14, 2013.

153. "Hydra," DARPA website, http://www.darpa.mil/Our_Work/TTO/Programs/Hydra.aspx.

154. Ray Locker, "Pentagon Developing Sub to Launch Air, Underwater Drones," *USA Today,* October 2, 2013, www.usatoday.com.

155. "DARPA's New Tern Program Aims For Eyes in the Sky," DARPA website, March 1, 2013, www.darpa.mil.

156. Ibid.

157. Douglas Aircraft Company, *Preliminary Design of an Experimental World-Circling Spaceship,* report no. SM-11827, May 2, 1946.

158. "Chronology of National Missile Defense Programs," Council on Foreign Relations website, June 1, 2002, www.cfr.org.

159. Johnson, *The Sorrows of Empire,* 81.

160. Ronald Reagan, "Address to the Nation on Defense and National Security" (speech, Oval Office, March 23, 1983), Ronald Reagan Presidential Library, www.reagan.utexas.edu.

161. Chalmers Johnson, *Nemesis: The Last Days of the American Republic* (New York: Henry Holt, 2007), 211.

162. William D. Hartung and Michelle Ciarrocca, "Push for Missile System Ignores Lack of a Real Threat," *Deseret News,* January 28, 2001, www.deseretnews.com.

163. Johnson, *The Sorrows of Empire,* 212.

164. *Report of the Commission to Assess the Ballistic Missile Threat to the United States,* executive summary, pursuant to Public Law 201, 104th Congress, Washington, D.C., July 15, 1998.

165. Hartung and Ciarrocca, "Push for Missile System Ignores Lack of a Real Threat."

166. Eric Schmitt, "Panel Says U.S. Faces Risk of a Surprise Missile Attack," *New York Times,* July 16, 1998, www.nytimes.com.

167. National Missile Defense Act of 1999, Public Law 106–38, 106th Congress, July 22, 1999, www.gpo.gov.

168. National Security Presidential Directive/NSPD-23, December 16, 2002, http://fas.org/irp/offdocs/nspd/nspd-23.htm.

169. *Report of the Commission to Assess United States National Security Space Management and Organization,* executive summary, pursuant to Public Law 106-65, Washington, D.C., January 11, 2001.

170. Johnson, *Nemesis,* 210.

171. U.S. Space Command, *Vision for 2020* (Peterson AFB, Colo.: U.S. Space Command, 1996).

172. William L. Shelton, "The Value of Space to the Warfighter" (speech, Air Force Association Mitchell Institute Friday Space Group Forum, Washington, D.C., February 7, 2014), Air Force Space Command, www.afspc.af.mil.

173. White House, *National Space Policy of the United States of America,* June 28, 2010, 3.

174. Department of Defense and the Office of the Director of National Intelligence, *National Security Space Strategy Unclassified Summary,* January 2011, 11.

175. Bruce Berkowitz, *The National Reconnaissance Office at 50 Years: A Brief History* (Chantilly, Va.: Center for the Study of National Reconnaissance, 2011).

176. Barton Gellman and Greg Miller, "'Black Budget' Summary Details U.S. Spy Network's Successes, Failures and Objectives," *Washington Post,* August 29, 2013, www.washingtonpost.com.

177. National Reconnaissance Program, *FY 2010 Congressional Budget Justification,* vol. 4, May 2009, 1.

178. Berkowitz, *The National Reconnaissance Office at 50 Years,* 11.

179. Steve Schofield, *Lifting the Lid on Menwith Hill: The Strategic Roles and Economic Impact of the US Spy Base In Yorkshire,* Yorkshire Campaign for Nuclear Disarmament (Nottingham, U.K.: Russell Press, 2012).

180. Mark Corcoran, "Drone Strikes Based on Work at Pine Gap Could See Australians Charged, Malcolm Fraser Says," *ABC News,* April 29, 2014, www.abc.net.au.

181. Alan Yuhas, "X-37B Secret Space Plane's Mission Remains Mystery outside U.S. Military," *Guardian,* October 27, 2014.

182. Alfred W. McCoy, "Imperial Illusions: Information Infrastructure and the Future of U.S. Global Power," in *Endless Empire: Spain's Retreat, Europe's Eclipse, America's Decline,* ed. Alfred W. McCoy, Josep M. Fradera, and Stephen Jacobson (Madison: University of Wisconsin Press, 2012), 375.

183. Ibid., 384.

4. The Rule by Nobody

1. Ulrich Beck, *World at Risk,* trans. Ciaran Cronin (Cambridge, U.K.: Polity Press, 2009), 10.

2. Ibid., 9.

3. Giorgio Agamben, *State of Exception,* trans. Kevin Attel (Chicago: University of Chicago, 2005).

4. Hannah Arendt, *The Human Condition,* 2nd ed. (Chicago: University of Chicago Press, 2013), 9.

5. Peter Sloterdijk, "Foreword to the Theory of Spheres," in *Cosmograms,* ed. Melik Ohanian and Jean-Christophe Royoux (New York: Lukas and Stenberg, 2005), 226.

6. Ibid.

7. Catherine Crump and Matthew Harwood, "Invasion of the Data Snatchers," *TomDispatch,* March 25, 2014, www.tomdispatch.com.

8. Paul J. Crutzen and Christian Schwägerl, "Living in the Anthropocene: Toward a New Global Ethos," *Yale Environment 306,* January 24, 2011, e360.yale.edu.

9. Nick Srnicek, "Representing Complexity: The Material Construction of World Politics" (PhD diss., London School of Economics, 2013), 255.

10. Senator Frank Church speaking on *Meet the Press,* NBC, August 17, 1975.

11. Glenn Greenwald, *No Place to Hide: Edward Snowden, The NSA, and the U.S. Surveillance State* (New York: Metropolitan Books, 2014), 169.

12. Michel Foucault, *"Society Must Be Defended": Lectures at the Collège de France, 1975–1976,* trans. David Macey (London: Penguin, 2003), 240.

13. Ibid., 249.

14. Roberto Esposito, *Bíos: Biopolitics and Philosophy,* trans. Timothy Campbell (Minneapolis: University of Minnesota Press, 2008), 46.

15. Ibid., 54–55.

16. Ibid., 54.

17. Ibid., 55.

18. Ibid., 58.

19. Alex Jeffrey, Colin McFarlane, and Alex Vasudevan, "Rethinking Enclosure: Space, Subjectivity, and the Commons," *Antipode* 44 (2012): 1259.

20. Center for Army Lessons Learned, *Commanders Guide to Biometrics in Afghanistan: Observations, Insights, and Lessons,* handbook no. 11-25 (Fort Leavenworth, Kans.: Center for Army Lessons Learned, 2011), 4.

21. Ibid., 5.

22. John O. Brennan, "The Ethics and Efficacy of the President's Counterterrorism Strategy" (remarks given at the Woodrow Wilson International Center for Scholars, Washington, D.C., April 30, 2012).

23. Gopal Ratnam, "Al-Qaeda 'Cancer' Spreads with U.S. Chasing, Panetta Says," *Bloomberg,* November 21, 2012, www.bloomberg.com.

24. Michael Hardt and Antonio Negri, *Empire* (Cambridge, Mass.: Harvard University Press, 2000), 189.

25. Esposito, *Bíos,* 14.

26. Eric H. Holder to Patrick J. Leahy, May 22, 2013.

27. Bernard Stiegler, *The Decadence of Industrial Democracies: Disbelief and Discredit,* vol. 1, trans. Daniel Ross and Suzanne Arnold (Cambridge, U.K.: Polity Press, 2011), 51.

28. Hannah Arendt, *On Violence* (New York: Houghton Mifflin Harcourt, 1969), 17.

29. Franz Kafka, *The Trial,* trans. John William (Ware, U.K.: Wordswith Edition, 2008), 3.

30. Max Weber, *Economy and Society: An Outline of Interpretive Sociology,* ed. Guenther Roth and Claus Wittich, trans. Ephraim Fischoff et al., 2nd ed. (Berkeley: University of California Press, 1978), 957.

31. Ibid.

32. Max Weber, *The Protestant Ethic and the Spirit of Capitalism,* trans. Stephen Kalberg (New York: Oxford University Press, 2011), 177.

33. Ibid.

34. Weber, *Economy and Society,* 981.

35. Arendt, *On Violence,* 81.

36. Ibid., 38.

37. Weber, *Economy and Society,* 987–88.

38. Chalmers Johnson, *Nemesis: The Last Days of the American Republic* (New York: Henry Holt and Company, 2007), 21.

39. Arendt, *The Human Condition,* 3.

40. Greg Miller, "Plan for Hunting Terrorists Signals U.S. Intends to Keep Adding Names to Kill Lists," *Washington Post,* October 23, 2012, www.washingtonpost.com.

41. Gregory S. McNeal. "Targeted Killing and Accountability," *Georgetown Law Journal* 102 (2014): 681–794.

42. Karen DeYoung, "CIA Veteran John Brennan Has Transformed U.S. Counterterrorism Policy," *Washington Post,* October 24, 2012, www.thewashingtonpost.com.

43. Jo Becker and Scott Shane, "Secret 'Kill List' Proves a Test of Obama's Principles and Will," *New York Times,* May 29, 2012, www.nytimes.com.

44. McNeal, "Targeted Killings and Accountability," 741.

45. David Kennedy, *Of War and Law* (Princeton, N.J.: Princeton University Press, 2006), 169.

46. Miller, "Plan for Hunting Terrorists."

47. Daniel Klaidman, *Kill or Capture: The War on Terror and the Soul of the Obama Presidency* (New York: Houghton Mifflin Harcourt, 2012), 202.

48. Michel Foucault, *Discipline and Punish: The Birth of the Prison,* trans. Alan Sheridan (London: Penguin Books, 1977), 201.

49. Gilles Deleuze, "Postscript on the Societies of Control," *October* 59 (1992): 3.

50. Foucault, *Discipline and Punish,* 308.

51. Deleuze, "Postscript on the Societies of Control," 4.

52. Ibid., 7.

53. Hardt and Negri, *Empire,* 197–98.

54. Martin Dodge and Rob Kitchin, "Code and the Transduction of Space," *Annals of the Association of American Geographers* 95, no. 1 (2005): 172.

55. Rob Kitchin and Martin Dodge, *Code/Space: Software and Everyday Life* (Cambridge, Mass.: MIT Press, 2011), 10.

56. Alain Badiou, *Logics of Worlds: Being and Event II,* trans. Alberto Toscano (London: Continuum, 2009), 422.

57. Arendt, *The Human Condition,* 73.

58. Ibid., 126.

59. Ibid., 147.

60. Bernard Stiegler, *Technics and Time,* vol. 1, *The Fault of Epimetheus,* trans. Richard Beardsworth and George Collins (Stanford, Calif.: Stanford University Press, 1998), 50.

61. Ibid., 177.

62. Arendt, *The Human Condition,* 147.

63. Ibid., 151.

64. Ibid., 322.

65. Ibid., 322–23.

66. Ibid., 257.

67. Ibid., 251.

68. Peter Wagner and Leah Sakala, "Mass Incarceration: The Whole Pie," *Prison Policy Initiative,* March 12, 2014, www.prisonpolicy.org.

69. Katie Rose Quandt, "Charts: Kids Are Paying the Price for America's Prison Binge," *Mother Jones,* August 26, 2014, www.motherjones.com.

70. Derek Thompson, "What Jobs Will the Robots Take?," *Atlantic,* January 23, 2014, www.theatlantic.com.

71. Bernard Stiegler, *Uncontrollable Societies of Disaffected Individuals: Disbelief and Discredit,* vol. 2, trans. Daniel Ross (Cambridge, U.K.: Polity Press, 2012), 16.

72. Ibid., 47.

73. Ibid., 88.

74. Edward Bernays, *Propaganda* (Brooklyn, N.Y.: Ig Publishing, 2005), 39–40.

75. Stiegler, *Uncontrollable Societies of Disaffected Individuals,* 121.

76. Stiegler, *The Decadence of Industrial Democracies,* 35.

77. Ibid., 49.

78. Stiegler, *Uncontrollable Societies of Disaffected Individuals,* 9.

79. Stiegler, *The Decadence of Industrial Democracies,* 37.

80. Arendt, *On Violence,* 86.
81. Ibid., 42.
82. Ibid., 50.
83. Anne Applebaum, *Gulag: A History of the Soviet Camps* (London: Penguin Books, 2003), 14.
84. Hannah Arendt, *The Origins of Totalitarianism* (New York: Houghton Mifflin Harcourt, 1973), 323.
85. Ibid, 438.
86. Ibid., 443.
87. Ibid., 456.
88. Hannah Arendt, *Eichmann and the Holocaust* (London: Penguin Books, 1963), 103.
89. Ibid., 60.
90. Ibid., 90.
91. Hannah Arendt, *The Life of the Mind* (New York: Houghton Mifflin Harcourt, 1971), 5.
92. Ibid., 176.
93. Ibid., 197.
94. Alain Badiou, *Metapolitics,* trans. Jason Barker (London: Verso, 2005), 7.
95. Ibid., 32.
96. Arendt, *The Life of the Mind,* 183.
97. Ibid., 177.
98. Peter W. Singer and Allan Friedman, *Cybersecurity and Cyberwar: What Everyone Needs to Know* (Oxford: Oxford University Press, 2014), 97.
99. Ibid., 69.
100. Richard A. Clarke and Robert K. Knake, *Cyber War: The Next Threat to National Security and What to Do about It* (New York: HarperCollins, 2010), 59.
101. Barton Gellman and Ellen Nakashima, "U.S. Spy Agencies Mounted 231 Offensive Cyber-Operations in 2011, Documents Show," *Washington Post,* August 30, 2013, www.washingtonpost.com.
102. "U.S. Needs 'Digital Warfare Force,'" BBC News, May 5, 2009, news.bbc.co.uk.
103. White House, *Cyberspace Policy Review: Assuring a Trusted and Resilient Information and Communications Structure* (Washington, D.C.: May 29, 2009), iii.
104. Singer and Friedman, *Cybersecurity and Cyberwar,* 2.
105. Ibid., 144.
106. Sanja Kelly et al., eds., *Freedom on the Net 2013: A Global Assessment of Internet and Digital Media* (Washington, D.C.: Freedom House, 2013).
107. Clarke and Knake, *Cyber War,* 63.
108. Ibid., 20.
109. David E. Sanger, David Barboza, and Nicole Perlroth, "Chinese Army

Unit Is Seen as Tied to Hacking against U.S.," *New York Times,* February 18, 2013, www.nytimes.com.

110. Clarke and Knake, *Cyber War,* 9.

111. John Arquilla and David Ronfeldt, "Cyberwar Is Coming!," *Comparative Strategy* 12, no. 2 (1993): 155.

112. Ben Quinn, "Revealed: The MOD's Secret Cyberwarfare Programme," *Guardian,* March 16, 2014, www.theguardian.com.

113. Nafeez Ahmed, "Pentagon Preparing for Mass Civil Breakdown," *Guardian,* June 12, 2014, www.theguardian.com.

114. James Bamford, "The Secret War," *Wired,* June 16, 2013, www.wired.com.

115. Select Committee to Study Governmental Operations with Respect to Intelligence Activities, United States Senate, *Supplementary Detailed Staff Reports on Intelligence Activities and the Rights of Americans,* bk. 3, report no. 94-755 (Washington, D.C.: U.S. Government Printing Office, April 23, 1976), 3.

116. Ibid., 765.

117. Ibid., 775.

118. Ibid., 751.

119. Greenwald, *No Place to Hide,* 128.

120. Edward C. Liu, *Amendments to the Foreign Intelligence Surveillance Act (FISA) Extended until June 1, 2015* (Washington, D.C.: Congressional Research Service, June 16, 2011).

121. United States Foreign Intelligence Surveillance Court, *In Re of the Federal Bureau of Investigation for an Order Requiring the Production of Tangible Things from[REDACTED],* docket no. BR 13-109 (Washington, D.C.: October 11, 2013), 14.

122. James Risen and Eric Lichtblau, "Bush Lets U.S. Spy on Callers without Courts," *New York Times,* December 16, 2005, www.nytimes.com.

123. Greenwald, *No Place to Hide,* 98.

124. Ryan Devereaux, Glenn Greenwald, and Laura Poitras, "Data Pirates of the Caribbean: The NSA Is Recording Every Cell Phone Call in the Bahamas," *Intercept,* May 19, 2014, https://firstlook.org.

125. Ryan Gallagher, "How Secret Partners Expand NSA's Surveillance Dragnet," *Intercept,* June 19, 2013, https://firstlook.org.

126. David E. Sanger and Thom Shanker, "NSA Devises Radio Pathway into Computers," *New York Times,* January 14, 2014, www.nytimes.com.

127. Gellman and Nakashima, "U.S. Spy Agencies Mounted 231 Offensive Cyber-Operations in 2011, Documents Show."

128. Greenwald, *No Place to Hide,* 148.

129. Glenn Greenwald and Ewen MacAskill, "NSA Prism Program Taps in to User Data of Apple, Google and Others," *Guardian,* June 7, 2013, www.theguardian.com.

130. Ewen MacAskill, "NSA Paid Millions to Cover Prism Compliance

Costs for Tech Companies," *Guardian,* August 23, 2013, www.theguardian.com.

131. Barton Gellman and Ashkan Soltani, "NSA Infiltrates Links to Yahoo, Google Data Centers Worldwide, Snowden Documents Say," *Washington Post,* October 30, 2013, www.washingtonpost.com.

132. Glenn Greenwald, "XKeyscore: NSA Tool Collects 'Nearly Everything a User Does on the Internet,'" *Guardian,* July 31, 2013, www.theguardian.com.

133. Glenn Greenwald and Ewen MacAskill, "Boundless Informant: The NSA's Secret Tool to Track Global Surveillance Data," *Guardian,* June 11, 2013, www.theguardian.com.

134. Barton Gellman and Ashkan Soltani, "NSA Tracking Cellphone Locations Worldwide, Snowden Documents Show," *Washington Post,* December 4, 2013, www.washingtonpost.com.

135. James Ball and Spencer Ackerman, "NSA Loophole Allows Warrantless Search for U.S. Citizens' Emails and Phone Calls," *Guardian,* August 9, 2013, www.theguardian.com.

136. Ryan Gallagher, "The Surveillance Engine: How the NSA Built Its Own Secret Google," *Intercept,* August 25, 2014, https://firstlook.org/.

137. John Shiffman and Kristina Cooke, "Exclusive: U.S. Directs Agents to Cover Up Program Used to Investigate Americans," Reuters, August 5, 2013, www.reuters.com.

138. Gallagher, "The Surveillance Engine."

139. Spencer Ackerman and James Ball, "Optic Nerve: Millions of Yahoo Webcam Images Intercepted by GCHQ," *Guardian,* February 28, 2014, www.theguardian.com.

140. United Nations, *Report of the Special Rapporteur on the Promotion and Protection of Human Rights and Fundamental Freedoms while Countering Terrorism* (New York: United Nations, 2014), 19.

141. Dana Priest, "NSA Growth Fueled by Need to Target Terrorists," *Washington Post,* July 21, 2013, www.washingtonpost.com.

142. Jeremy Scahill and Glenn Greenwald, "The NSA's Secret Role in the U.S. Assassination Program," *Intercept,* February 10, 2014, https://firstlook.org.

143. Ibid.

144. Ibid.

145. Ibid.

146. Greg Miller, Julie Tate, and Barton Gellman, "Documents Reveal NSA's Extensive Involvement in Targeted Killing Program," *Washington Post,* October 16, 2013, www.washingtonpost.com.

147. Scahill and Greenwald, "The NSA's Secret Role in the U.S. Assassination Program."

148. Tom Engelhardt, "Tomgram: It's about Blackmail, Not National Security," *TomDispatch,* January 19, 2014, www.tomdispatch.com.

149. Greenwald, *No Place to Hide,* 47–48.

150. "Interview with Bernard Stiegler," *French Culture,* October 3, 2013, www.frenchculture.org.

151. Jacques Rancière, *Disagreement: Politics and Philosophy,* trans. Julie Rose (Minneapolis: University of Minnesota Press, 2004), 29.

152. Alfred McCoy, "Surveillance and Scandal," *TomDispatch,* January 19, 2014, www.tomdispatch.com.

153. Hardt and Negri, *Empire,* 23.

154. Naomi Wolf, "Fascist America, in 10 Easy Steps," *Guardian,* April 24, 2007, www.theguardian.com.

155. Greenwald, *No Place to Hide,* 173.

156. Matthew Cole, Richard Esposito, Bill Dedman, and Mark Schone, "Inside the Mind of Edward Snowden," *ABC News,* May 29, 2014, www.nbc .com

157. Gilles Deleuze and Félix Guattari, *A Thousand Plateaus: Capitalism and Schizophrenia,* trans. Brian Massumi (London: Continuum, 2004), 415.

158. James Bamford, "The Agency That Could Be Big Brother," *New York Times,* December 25, 2005.

159. Senator Frank Church speaking on *Meet the Press,* NBC, August 17, 1975.

160. Arendt, *The Life of the Mind,* 193.

5. Policing Everything

1. Alfred W. McCoy, *Policing America's Empire: The United States, the Philippines, and the Rise of the Surveillance State* (Madison: University of Wisconsin Press, 2009).

2. Ibid., 13.

3. Ibid., 17.

4. Ibid., 8.

5. Michael Hardt and Antonio Negri, *Empire* (Cambridge, Mass.: Harvard University Press, 2000), xv.

6. Michel Foucault, *Discipline and Punish: The Birth of the Prison,* trans. Alan Sheridan (London: Penguin Books, 1977), 304.

7. Michel Foucault, *"Society Must Be Defended": Lectures at the Collège de France, 1975–1976,* trans. David Macey (London: Penguin, 2003), 46–47.

8. Ibid., 109.

9. Mike Davis, "The Urbanization of Empire: Megacities and the Laws of Chaos," *Social Text* 22, no. 4 (2004): 14.

10. Mark Neocleous, *War Power, Police Power* (Edinburgh: Edinburgh University Press, 2014), 137.

11. E. P. Thompson, *The Making of the English Working Class* (Harmondsworth, U.K.: Penguin Books, 1963), 242.

12. E. P. Thompson, *Customs in Common* (London: Penguin Books, 1991), 184.

13. Neocleous, *War Power, Police Power,* 66.
14. Ibid., 6.
15. Karl Marx, *Capital: A Critique of Political Economy,* trans. Ben Fowkes (London: Penguin Books, 1990), 873.
16. Ibid., 875.
17. Peter Linebaugh, *Stop, Thief! The Commons, Enclosures, and Resistance* (Oakland, Calif.: PM Press, 2014), 13.
18. Marx, *Capital,* 875.
19. Linebaugh, *Stop, Thief!,* 32.
20. Thompson, *Customs in Common,* 102.
21. "Enclosing the Land," U.K. Parliament website, www.parliament.uk.
22. Linebaugh, *Stop, Thief!,* 80.
23. Marx, *Capital,* 878.
24. Thompson, *Customs in Common,* 122.
25. Ibid., 125.
26. Hardt and Negri, *Empire,* 326.
27. Thompson, *Customs in Common,* 163.
28. Peter Linebaugh, *Stop, Thief! The Commons, Enclosures, and Resistance* (Oakland, Calif.: PM Press, 2014), 8.
29. Foucault, *Discipline and Punish,* 85.
30. Thompson, *The Making of the English Working Class,* 238.
31. Thompson, *Customs in Common,* 176.
32. Neocleous, *War Power, Police Power,* 59.
33. Foucault, *"Society Must Be Defended,"* 50.
34. Friedrich Engels, *The Condition of the Working Class in England,* trans. Tristram Hunt (London: Penguin Books, 2009), 69.
35. Karl Marx, "The June Revolution: June 29, 1848," trans. Marx-Engels Institute, Marxists Internet Archive, www.marxists.org.
36. Marx, *Capital,* 895.
37. Neocleous, *War Power, Police Power,* 30.
38. Marx, *Capital,* 896.
39. Ibid., 899.
40. Neocleous, *War Power, Police Power,* 32.
41. Foucault, *Discipline and Punish,* 87.
42. Ibid., 130.
43. Ibid., 141.
44. Linebaugh, *Stop, Thief!,* 1.
45. Foucault, *Discipline and Punish,* 228.
46. Ibid., 169.
47. Ibid., 148.
48. Ibid., 249.
49. Ibid., 201.
50. Ibid., 216.
51. Hardt and Negri, *Empire,* 329.

52. Stephen Graham, *Cities under Siege: The New Military Urbanism* (London: Verso, 2010), 146.

53. Linebaugh, *Stop, Thief!,* 38.

54. Foucault, *"Society Must Be Defended,"* 245.

55. David Harvey, "The Right to the City," *New Left Review* 53 (2008): 33.

56. Ibid., 24.

57. Neil Smith, *The New Urban Frontier: Gentrification and the Revanchist City* (London: Routledge, 1996), 220.

58. Linebaugh, *Stop, Thief!,* 157.

59. Eyal Weizman, "Political Plastic," *Collapse* 6 (2010): 261.

60. Ibid., 268.

61. Steve Herbert, *Policing Space: Territoriality and the Los Angeles Police Department,* (Minneapolis: University of Minnesota Press, 1997), 13.

62. Weizman, "Political Plastic," 275.

63. Joe Painter, "Rethinking Territory," *Antipode* 42, no. 5 (2010): 1090–118.

64. Rob Kitchin and Martin Dodge, "Code and the Transduction of Space," *Annals of the Association of American Geographers* 95, no. 1 (2005): 162–80.

65. Radley Balko, *Rise of the Warrior Cop: The Militarization of America's Police Forces* (New York: PublicAffairs, 2013), 14.

66. Ibid., 23–25.

67. Herbert, *Policing Space,* 60.

68. Balko, *Rise of the Warrior Cop,* 34–35.

69. Herbert, *Policing Space,* 81.

70. Ibid., 149.

71. Ibid.

72. Balko, *Rise of the Warrior Cop,* 52.

73. Richard Nixon, "Special Message to the Congress on Drug Abuse Prevention and Control," June 17, 1971, posted by Gerhard Peters and John T. Woolley, American Presidency Project, www.presidency.ucsb.edu/ws/?pid=3048.

74. Emily Dufton, "The War on Drugs: How President Nixon Tied Addiction to Crime," *Atlantic,* March 26, 2012, www.theatlantic.com.

75. Ibid.

76. Balko, *Rise of the Warrior Cop,* 139.

77. Ibid., 158.

78. Ibid.

79. American Civil Liberties Union, *War Comes Home: The Excessive Militarization of American Policing* (New York: American Civil Liberties Union, June 2014), 2.

80. "Drug War Statistics," Drug Policy Alliance, www.drugpolicy.org.

81. Foucault, *"Society Must Be Defended,"* 50–51.

82. Matt Apuzzo, "War Gear Flows to Police Departments," *New York Times,* June 8, 2014, www.nytimes.com.

83. Law Enforcement Support Office, "About the 1033 Program," www.dispositionservices.dla.mil.

84. Balko, *Rise of the Warrior Cop,* 188.

85. Ibid., 180.

86. Faiza Patel and Michael Price, "Ferguson is Not Fallujah," *Just Security,* August 21, 2014, www.justsecurity.org.

87. Glenn Greenwald, "The Militarization of U.S. Police: Finally Dragged into the Light by the Horrors of Ferguson," *Intercept,* August 14, 2014, https://firstlook.org.

88. Balko, *Rise of the Warrior Cop,* 308.

89. American Civil Liberties Union, *War Comes Home,* 18.

90. Joseph Nevins, *Operation Gatekeeper and Beyond: The War on "Illegals" and the Remaking of the U.S.–Mexico Boundary* (New York: Routledge, 2010).

91. Geoffrey Boyce and Jill Williams, "Intervention: Homeland Security and the Precarity of Life in the Borderlands," *Antipode,* December 10, 2012, www.antipodefoundation.org.

92. Todd Miller and Gabriel M. Schivone, "Gaza in Arizona," *TomDispatch,* January 25, 2015.

93. Armando Carrasco and Second Lieutenant Vincent Borrello, "On the Border: Unit Volunteers for JTF North Support Mission," *Fort Bliss Monitor,* April 7, 2011, www.jtfn.northcom.mil.

94. Boyce and Williams, "Homeland Security."

95. Todd Miller, "They Are Watching You," *TomDispatch,* April 22, 2014, www.tomdispatch.com.

96. Aleem Maqbool, "African-American Police Officer: Ferguson 'Heart Wrenching,'" BBC, August 25, 2014, www.bbc.co.uk/news.

97. Apuzzo, "War Gear Flows to Police Departments."

98. Spencer Ackerman, "U.S. Police Given Billions from Homeland Security for 'Tactical' Equipment," *Guardian,* August 20, 2014, www.theguardian.com.

99. Taylor Wofford, "How America's Police Became an Army: The 1033 Program," *Newsweek,* August 13, 2014, www.newsweek.com.

100. American Civil Liberties Union, *War Comes Home,* 22.

101. Ibid., 4.

102. Greenwald, "The Militarization of U.S. Police."

103. Balko, *Rise of the Warrior Cop,* 13.

104. Gilbert Mercier, "The U.S. War Culture Has Come Home to Roost," *CounterPunch,* August 20, 2014, www.counterpunch.org.

105. Tamara K. Nopper and Mariame Kaba, "Itemizing Atrocity," *Jacobin,* August 15, 2014, www.jacobinmag.com.

106. U.S. Department of Justice, *Investigation of the Ferguson Police Department* (Washington, D.C.: U.S. Department of Justice, 2015), 2.

107. Ibid.

108. Amanda Holpuch, "FBI Director Says Racism Not Epidemic in Police But Is 'Cultural Inheritance' of U.S.," *Guardian,* February 12, 2015.

109. "Justifiable Homicide, by Weapon, Law Enforcement, 2009–2013," FBI website, www.fbi.gov.

110. Tom McCarthy, "The Uncounted: Why the U.S. Can't Keep Track of People Killed by Police," *Guardian,* March 18, 2015, www.theguardian.com.

111. Kara Dansky, "The Real Reason Ferguson Has Military Weapons," CNN, August 19, 2014, www.cnn.com.

112. U.S. Government Accountability Office, *Additional Actions Could Help Ensure That Efforts to Share Terrorism-Related Suspicious Activity Reports Are Effective* (Washington, D.C.: U.S. Government Accountability Office, 2013).

113. Hina Shamsi and Matthew Harwood, "Uncle Sam's Databases of Suspicion," *TomDispatch,* November 6, 2014, www.theguardian.com.

114. Jeremy Scahill and Ryan Devereaux, "Barack Obama's Secret Terrorist-Tracking System, by the Numbers," *Intercept,* August 5, 2014.

115. McCoy, *Policing America's Empire.*

116. Ibid., 24.

117. Executive Office of the President, National Science and Technology Council Subcommittee on Biometrics and Identity Management, *The National Biometrics Challenge* (Washington D.C.: National Science and Technology Council, September 2011), 5.

118. Jennifer Lynch, "FBI Plans to Have 52 Million Photos in Its NGI Face Recognition Database by Next Year," Electronic Frontier Foundation, April 14, 2014, www.eff.org.

119. "Memphis PD: Keeping Ahead of Criminals by Finding the 'Hot Spots,'" IBM, 2011, www.ibm.com.

120. Matt Stroud, "The Minority Report: Chicago's New Police Computer Predicts Crimes, But Is It Racist?," *Verge,* February 19, 2014, www.theverge.com.

121. Patrick Healy, "Predictive Policing Forecasts Crime That Officers Then Try to Deter," NBCLA, January 8, 2013, www.nbclosangeles.com.

122. Nate Berg, "Predicting Crime, LAPD-Style," *Guardian,* June 25, 2014, www.theguardian.com.

123. Stroud, "The Minority Report."

124. Jeremy Gorner, "Chicago Police Use 'Heat List' as Strategy to Prevent Violence," *Chicago Tribune,* August 21, 2013, articles.chicagotribune.com.

125. Jon Campbell, "License Plate Recognition Logs Our Lives Long before We Sin," *LA Weekly,* June 21, 2012, www.laweekly.com.

126. Ibid.

127. Conor Friedersdorf, "Eyes over Compton: How Police Spied on a Whole City," *Atlantic,* April 21, 2014.

128. Louise Amoore, Stephen Marmura, and Mark B. Salter, "Editorial: Smart Borders and Mobilities: Spaces, Zones, Enclosures," *Surveillance and Society* 5, no. 2 (2008): 96.

129. Ibid.

130. Mike Davis, *Planet of Slums* (London: Verso, 2006), 19.

131. Graham, *Cities under Siege,* 109.

132. Engels, *The Condition of the Working Class in England,* 69.

133. Martin Heidegger, *Being and Time,* trans. Joan Stambaugh (Albany: State University of New York Press, 2010).

134. Hubert L. Dreyfus, *Being-in-the-World: A Commentary on Heidegger's Being and Time, Division I* (Cambridge, Mass.: MIT Press, 1991), 5.

135. Heidegger, *Being and Time,* 108.

136. Peter Sloterdijk, "Foreword to the Theory of Spheres," in *Cosmograms,* ed. Melik Ohanian and Jean-Christophe Royoux (New York: Lukas and Stenberg, 2005), 234.

137. Ibid., 236.

138. Hardt and Negri, *Empire,* 188.

139. Allen Feldman, "Securocratic Wars of Public Safety," *Interventions* 6, no. 3 (2004): 335.

140. Sloterdijk, "Foreword to the Theory of Spheres," 239.

141. Ibid., 236.

142. Francisco R. Klauser, "Splintering Spheres of Security: Peter Sloterdijk and the Contemporary Fortress City," *Environment and Planning D: Society and Space* 28, no. 2 (2010): 331.

143. Ibid., 335.

144. Lieven de Cauter, *The Capsular Civilization: On the City in the Age of Fear* (Rotterdam, Netherlands: NAi Publishers, 2004),

145. De Cauter, *The Capsular Civilization,* 45.

146. Ibid., 45.

147. Don Mitchell, "The S.U.V. Model of Citizenship: Floating Bubbles, Buffer Zones, and the Rise of the 'Purely Atomic' Individual." *Political Geography* 24, no. 1 (2005): 96.

148. Graham, *Cities under Siege,* 316.

149. Ibid., 107.

150. Peter Sloterdijk, "Foam City: About Urban Spatial Multitudes," trans. Antonio Petrov, *New Geographies* 0 (2009): 137–38.

151. Stephen Graham and Lucy Hewitt, "Getting off the Ground: On the Politics of Urban Verticality," *Progress in Human Geography* 37, no. 1 (2013): 79.

152. De Cauter, *The Capsular Civilization,* 46.

153. Ben Anderson "Affective Atmospheres," *Emotion, Space, and Society* 2, no.2 (2009): 78.

154. Ibid., 80.
155. Mark Neocleous, *War Power, Police Power,* 206–7
156. Davis, "The Urbanization of Empire," 15.
157. Graham, *Cities under Siege,* 9.
158. De Cauter, *The Capsular Civilization,* 45.
159. Graham, *Cities under Siege.*
160. U.S. Army, *Military Operations on Urbanized Terrain (MOUT),* FM 90-10 (Washington, D.C.: U.S. Army, August 15, 1979).
161. Ralph Peters, "Our Soldiers, Their Cities," *Parameters* 26 (Spring 1996): 43–50, strategicstudiesinstitute.army.mil/pubs/parameters/Articles/96spring/peters.htm.
162. Graham, *Cities under Siege,* XVI.
163. Feldman, "Securocratic Wars of Public Safety."
164. Ibid., 331.
165. Graham, *Cities under Siege,* 94.
166. Feldman, "Securocratic Wars of Public Safety," 334.
167. Peter Adey, "Security Atmospheres or the Crystallisation of Worlds," *Environment and Planning D: Society and Space* 32, no. 5 (2014): 835.
168. Feldman, "Securocratic Wars," 340.
169. Eyal Weizman, "Maps of Israeli Settlements," *openDemocracy,* April 25, 2002, www.opendemocracy.net.
170. Stuart Elden, "Secure the Volume: Vertical Geopolitics and the Depth of Power," *Political Geography* 34 (2013): 35.
171. Eyal Weizman, "Control in the Air," *openDemocracy,* May 2, 2002, www.opendemocracy.net.
172. Sloterdijk, "Foreword to the Theory of Spheres," 225.
173. Department of the Army, *Civil Disturbances,* ATP 3–39.33 (Washington, D.C.: Department of the Army, April 21, 2014).
174. Anna Feigenbaum, "100 Years of Tear Gas," *Atlantic,* August 16, 2014, www.theatlantic.com.
175. Ibid.
176. N. A. J. Taylor, "Teargas: Or, the State as Atmo-Terrorist," *Al Jazeera,* May 5, 2012, www.aljazeera.com.
177. Marijn Nieuwenhuis, "Terror in the Air in Istanbul," *Society and Space Open Site,* June 20, 2013, www.societyandspace.com.
178. Peter Sloterdijk, *Terror from the Air,* trans. Amy Patton and Steve Corcoran (Los Angeles: Semiotext[e], 2009), 100.
179. Phillipe Theophanidis, "Caught in the Cloud: The Biopolitics of Tear Gas Warfare," *Funambulist,* August 26, 2013, www.thefunambulist.net.
180. International Human Rights and Conflict Resolution Clinic at Stanford Law School and Global Justice Clinic at NYU School of Law, *Living under Drones: Death, Injury, and Trauma to Civilians from U.S. Drone Practices in Pakistan* (September 2012), 151.

181. Craig Whitlock, "Close Encounters on Rise as Small Drones Gain in Popularity," *Washington Post,* June 23, 2014, www.washingtonpost.com.

182. Craig Whitlock, "FBI Has Received Aviation Clearance for at Least Four Domestic Drone Operations," *Washington Post,* June 20, 2013, www.thewashingtonpost.com.

183. U.S. Department of Justice, *Interim Report on the Department of Justice's Use and Support of Unmanned Aircraft System,* Report 13-37 (September 2013), 4.

184. Thomas A. Peter, "Drones on the U.S. Border: Are They Worth the Price?," *Christian Science Monitor,* February 5, 2014, www.csmonitor.com.

185. Jennifer Lynch, "Customs and Border Protection Loaned Predator Drones to Other Agencies 700 Times in Three Years According to 'Newly Discovered' Records," *Electronic Frontier Foundation,* January 14, 2014, www.eff.org.

186. Ibid.

187. Declan McCullagh, "DHS Built Domestic Surveillance Tech into Predator Drones," CNET, March 2, 2013, www.cnet.com.

188. Naomi Wolf, "The Coming Drone Attack on America," *Guardian,* December 21, 2012, www.theguardian.com.

189. U.S. Army UAS Center of Excellence, *U.S. Army Unmanned Aircraft Systems Roadmap 2010–2035* (Fort Rucker, Ala.: U.S. Army, 2010), 4.

190. Greenwald, "The Militarization of U.S. Police."

191. U.S. Government Accountability Office, *Defense Acquisitions: Assessments of Selected Weapon Program,* GAO-13-294SP (Washington, D.C.: U.S. Government Accountability Office, 2013), 143.

192. Craig Timberg, "Blimplike Surveillance Craft Set to Deploy over Maryland Heighten Privacy Concerns," *Washington Post,* January 22, 2014, www.washingtonpost.com.

193. Mark Neocleous, *War Power, Police Power,* 162.

194. Chalmers Johnson, *Blowback: The Costs and Consequences of American Empire,* 2nd ed. (London: Time Warner Paperbacks, 2002), 13.

195. Graham and Hewitt, "Getting off the Ground," 86.

196. Noah Shachtman, "Army Tracking Plan: Drones That Never Forget a Face," *Wired,* September 28, 2011, www.wired.com.

197. Ibid.

198. U.S. Department of Defense, *Unmanned Systems Integrated Roadmap, FY2013–2038,* 14-S-0553 (2013), 12–13.

199. Robert Barkan, "Nobody Here but Us Robots," *New Republic,* April 29, 1972, 15.

200. John Arquilla and David Ronfeldt, *Swarming and the Future of Conflict* (RAND National Defense Research Institute, 2000), vii.

201. U.S. Army, *U.S. Army Unmanned Aircraft Systems Roadmap 2010–2035,* 65.

202. Defense Advanced Research Projects Agency, "Operating in Contested Environments," DARPA, March 30, 2015, www.darpa.mil.

203. Alfred McCoy, "Fatal Florescence: Europe's Decolonization and America's Decline," in *Endless Empire: Spain's Retreat, Europe's Eclipse, America's Decline,* ed. Alfred W. McCoy, Josep M. Fradera, and Stephen Jacobson (Madison: University of Wisconsin Press, 2012), 34.

204. Derek Gregory, "The Everywhere War," *Geographical Journal* 177 (2011): 238–50.

205. Ariel Handel "Exclusionary Surveillance and Spatial Uncertainty in the Occupied Palestinian Territories," in *Israel/Palestine: Population, Territory, and Power,* ed. Elia Zureik, David Lyon, Yasmeen Abu-Laban (New York: Routledge, 2011), 259–75.

206. David Harvey, *Seventeen Contradictions and the End of Capitalism* (London: Profile Books, 2014), 292.

207. Neocleous, *War Power, Police Power,* 162.

Conclusion

1. Tom Engelhardt, "The Fourth Branch: The Rise to Power of the National Security State," *TomDispatch,* August 3, 2014, www.tomdispatch .com.

2. George W. Bush, "Transcript of President Bush's address to a joint session of Congress on Thursday night, September 20, 2001," CNN, September 21, 2001, http://edition.cnn.com.

3. George W. Bush, "Text of President Bush's Address to the Nation," *Washington Post,* September 11, 2006, www.washingtonpost.com.

4. Mark Neocleous, *War Power, Police Power* (Edinburgh: Edinburgh University Press, 2014), 137.

5. Michael Hardt and Antonio Negri, *Empire* (Cambridge, Mass.: Harvard University Press, 2000), xiv.

6. Andrew J. Bacevich, *Washington Rules: America's Path to Permanent War* (New York: Metropolitan Books, 2010).

7. "Sen. Burr: Terror Threat a 'War on Western Civilization,'" *RealClear Politics,* January 11, 2015, www.realclearpolitics.com.

8. Roberto Esposito, *Bíos: Biopolitics and Philosophy,* trans. Timothy Campbell (Minneapolis: University of Minnesota Press, 2008), 148.

9. Chalmers Johnson, *Blowback: The Costs and Consequences of American Empire,* 2nd ed. (London: Time Warner Paperbacks, 2002), 229.

10. Tom Engelhardt, *Shadow Government: Surveillance, Secret Wars, and a Global Security State in a Single-Superpower World* (Chicago: Haymarket Books, 2014).

11. Michel Foucault, *"Society Must Be Defended": Lectures at the Collège de France, 1975–1976,* trans. David Macey (London: Penguin, 2003), 29.

12. Bruno Latour and Michael Callon, "Unscrewing the Big Leviathan: How Actors Macro-structure Reality and How Sociologists Help Them to Do So," in *Advances in Social Theory and Methodology: Toward an Integration of Micro and Macro-sociologies,* ed. Karin Knorr-Cetina and Aaron Victor Cicourel (Boston: Routledge and Kegan Paul, 1981), 294.

13. Hannah Arendt, *On Violence* (New York: Houghton Mifflin Harcourt, 1969), 4.

14. Barack Obama, "Remarks by the President at the National Defense University" (speech, National Defense University, Washington, D.C., May 23, 2013).

15. Stephen Graham, *Cities under Siege: The New Military Urbanism* (London: Verso, 2010), 180.

16. Arendt, *On Violence,* 81.

17. Latour and Callon, "Unscrewing the Big Leviathan," 295.

18. Arendt, *On Violence,* 50.

19. Bruno Latour, *Politics of Nature: How to Bring the Sciences into Democracy,* trans. Catherine Porter (Cambridge, Mass.: Harvard University Press, 2004), 218.

20. Arendt, *On Violence,* 17.

21. Hannah Arendt, *The Human Condition,* 2nd ed. (Chicago: University of Chicago Press, 2013), 322–23.

22. Gabriella Blum, "The Individualization of War: From War to Policing in the Regulation of Armed Conflict," in *Law and War,* ed. Austin Sarat, Lawrence Douglas, and Martha Merrill Umphrey (Stanford, Calif.: Stanford University Press, 2014), 66.

23. Tom Engelhardt, "Remotely Piloted War," *TomDispatch,* February 23, 2012, www.tomdispatch.com.

24. Paul Dickson, *The Electronic Battlefield* (Takoma Park, Md.: FoxAcre Press, 2012), 208.

25. Arendt, *On Violence,* 42.

26. Obama, "Remarks by the President at the National Defense University."

27. Matthew Harwood, "The Lone-Wolf Terror Trap," *TomDispatch,* February 5, 2015, www.tomdispatch.com.

28. Mary Kaldor, *New and Old Wars: Organized Violence in a Global Era,* 3rd ed. (Cambridge, U.K.: Polity Press, 2012), 6–7.

29. Blum, "The Individualization of War," 56.

30. Grégoire Chamyou, *Drone Theory,* trans. Janet Loyd (New York: Penguin, 2015), 53.

31. Stuart Elden, *Terror and Territory: The Spatial Extent of Sovereignty* (Minneapolis: University of Minnesota Press. 2009), 173.

32. Hardt and Negri, *Empire,* 189.

33. Derek Gregory, "The Everywhere War," *Geographical Journal* 177, no. 3 (2011): 238–50.

34. Jenifer Rubin, "Sen. Lindsey Graham: Boston Bombing Is Exhibit A of Why the Homeland Is the Battlefield," *Washington Post,* April 19, 2013, www.washingtonpost.com.

35. Adam Serwer, "Obama Administration Says President Can Use Lethal Force against Americans on U.S. Soil," *Mother Jones,* March 5, 2013, www.motherjones.com.

36. Harwood, "The Lone-Wolf Terror Trap."

37. Bernard Stiegler, *Uncontrollable Societies of Disaffected Individuals: Disbelief and Discredit,* vol. 2, trans. Daniel Ross (Cambridge, U.K.: Polity Press, 2012), 47.

38. Alain Badiou, *Infinite Thought and the Return to Philosophy,* trans. Oliver Feltham and Justin Clemens (London: Continuum, 2005), 121.

39. Peter Sloterdijk, *In the World Interior of Capital: Towards a Philosophical Theory of Globalization,* trans. Wieland Hoban (Cambridge, U.K.: Polity Press, 2013), 30.

40. Ibid., 140.

41. Friedrich Engels, "Speeches in Elberfeld," Marx and Engels Internet Archive, February 8, 1845, https://www.marxists.org/.

42. Peter Sloterdijk, "Foreword to the Theory of Spheres," in *Cosmograms,* ed. Melik Ohanian and Jean-Christophe Royoux (New York: Lukas and Stenberg, 2005), 239.

43. Hardt and Negri, *Empire,* 337.

44. Peter Sloterdijk, *Spheres,* vol. 1, *Bubbles,* trans. Wieland Hoban (Los Angeles: Semiotext[e], 2011), 71.

45. E. P. Thompson, *The Making of the English Working Class* (Harmondsworth, U.K.: Penguin Books, 1963), 239.

46. Ibid., 238.

47. Hardt and Negri, *Empire,* 339.

48. Esposito, *Bíos: Biopolitics and Philosophy,* 55.

49. Jacques Ellul, *The Technological Society,* trans. John Wilkinson (New York: Vintage Books, 1964), 117.

50. Ariel Handel, "Exclusionary Surveillance and Spatial Uncertainty in the Occupied Palestinian Territories," in *Israel/Palestine: Population, Territory and Power,* ed. Elia Zureik, David Lyon, Yasmeen Abu-Laban (New York: Routledge, 2011), 259–75.

51. Achille Mbembe, "Necropolitics," *Public Culture* 15, no. 1 (2003): 40.

52. Engelhardt, *Shadow Government,* 97.

53. Eyal Weizman, "Thanatotactics," *Springerin,* April 2006, http://www.springerin.at/.

54. Paul Virilio, *War and Cinema: The Logistics of Perception,* trans. Patrick Camiller. (London: Verso, 1989), 59.

55. Sloterdijk, *In the World Interior of Capital,* 139.

56. Virilio, *War and Cinema,* 8.

57. Ibid, 91.

58. Guy Debord, *Society of the Spectacle,* trans. Fredy Perlman et al., rev. ed. (Kalamazoo, Mich.: Black and Red, 1977), https://www.marxists.org/reference/archive/debord/society.htm.

59. Giorgio Agamben, *What Is an Apparatus? And Other Essays,* trans. David Kishik and Stefan Pedatella (Stanford, Calif.: Stanford University Press, 2009), 14.

60. Ibid, 20.

61. International Human Rights and Conflict Resolution Clinic at Stanford Law School and Global Justice Clinic at NYU School of Law, *Living under Drones: Death, Injury, and Trauma to Civilians from U.S. Drone Practices in Pakistan* (September 2012), 98.

62. Chamayou, *Drone Theory,* 45.

63. Arendt, *The Human Condition,* 257.

64. Hannah Arendt, *The Origins of Totalitarianism* (New York: Houghton Mifflin Harcourt, 1968), 438.

65. Bernard Stiegler, *The Decadence of Industrial Democracies: Disbelief and Discredit,* vol. 1, trans. Daniel Ross and Suzanne Arnold (Cambridge, U.K.: Polity Press, 2011), 51.

66. Arendt, *On Violence,* 50.

67. Stiegler, *The Decadence of Industrial Democracies,* 50.

68. Arendt, *The Human Condition,* 147.

69. Thomas Hobbes, *Leviathan: With Selected Variants from the Latin Edition of 1668,* ed. Edwin Curley (Indianapolis: Hackett Publishing Company, 1994), 106.

70. Michel Foucault, *Discipline and Punish: The Birth of the Prison,* trans. Alan Sheridan (London: Penguin Books, 1977), 217.

71. Ibid., 169.

72. Joseph A. Meyer, "Crime Deterrent Transponder System," *IEEE Transactions of Aerospace and Electronic Systems* 7, no. 1 (1971): 17.

73. Sloterdijk, *Spheres,* 25.

74. Associated Press, "Vermont Town's Dome Dream Dims Like Words to Anthem," *Toledo Blade,* October 12, 1981.

Index

Ian G. R. Shaw is lecturer in human geography at the University of Glasgow and author of numerous articles on drone warfare and political philosophy.